MOLECULAR BIOLOGY AND BIOTECHNOLOGY

Special Publication No 54

Molecular Biology and Biotechnology

Based on Lectures given at a Residential School,
Organised by the Post-Experience Courses Committee of
The Royal Society of Chemistry

Hatfield Polytechnic, England, 15th–19th July 1985

Edited by
J. M. Walker
E. B. Gingold
The Hatfield Polytechnic

The Royal Society of Chemistry
Burlington House, London W1V 0BN

British Library Cataloguing in Publication Data
Molecular biology and biotechnology : based on lectures given at a residential school organised by the Post-Experience Courses Committee of the Royal Society of Chemistry, Hatfield Polytechnic, England, 15th–19th July 1985.—(Special publication, ISSN 0260–6291; no. 54)
1. Molecular biology
I. Walker, John M. II. Gingold, E. B. III. Royal Society of Chemistry IV. Series
574.8'8 QH506

ISBN 0-85186-985-8

Printed in Great Britain by Henry Ling Ltd., at the Dorset Press, Dorchester, Dorset

Preface

Had one been asked twenty years ago to select the most theoretical and least applied areas of the biological sciences, there is no doubt that molecular biology would have been near the top of the list. Yet today this same area has become the centre of an unparalleled expansion of applied biology, for out of these theoretical studies has come genetic engineering, and with it the ability to manipulate organisms to produce products previously only obtainable by more difficult and expensive routes. So dramatic has been the increase in possibilities that a new term, 'biotechnology', has been introduced to cover this field.

Of course, the use of biological processes to manufacture products, ranging from alcohol to antibiotics, is by no means new. Biotechnology may be a new term, but the science is firmly rooted in more traditional fields such as industrial microbiology and process biochemistry. However, the recent developments in molecular biology are without doubt responsible for the enormous public interest in biotechnology. The organisation of a Royal Society of Chemistry Residential Course on this topic at Hatfield Polytechnic in July 1985, from which many of the contributions in this book are drawn, is one reflection of this interest. The theme of the course, and of the book, is that the advances in biotechnology should be viewed in the context of the long standing industrial biology. For this reason the book begins with a chapter reviewing the use of microbes for industrial processes. Following this there are two chapters introducing the methodology of genetic engineering, with particular emphasis on selection of microbes manipulated to produce industrially important products. However, genetic engineering has now moved beyond just manipulating bacteria, as the next three chapters on yeast, mammalian cells, and crop plants demonstrate.

The new technology has had many applications. We have included chapters on the production of pharmaceutical compounds, on clinical diagnosis and on the study of the agriculturally important question of the ripening of fruit, to give the reader an indication of the diversity of these applications. Genetic engineering is not the only area of molecular biology to have made a major contribution to industry, as the remaining chapters on enzyme technology and its applications testify. The contributions in this area should also give the reader some idea of the problems in converting laboratory processes to industrial scale.

The R.S.C. course was aimed not at expert biotechnologists but at scientific workers with little or no previous experience in this field. The contributions in this book should thus be seen as primarily having a teaching function. The book should prove of interest both to undergraduates studying for biological or chemical qualifications and to scientific workers from other fields who need a basic introduction to this rapidly expanding area.

J.M. Walker
E.B. Gingold

Contents

1
Products from Micro-organisms

By P. F. Stanbury

DIVISION OF BIOLOGICAL AND ENVIRONMENTAL SCIENCES, THE HATFIELD POLYTECHNIC, P.O. BOX 109, COLLEGE LANE, HATFIELD, HERTS. AL10 9AB, U.K.

Introduction

Micro-organisms are capable of growing on a wide range of substrates and can produce a remarkable spectrum of products. The relatively recent advent of *in vitro* genetic manipulation has extended the range of products that may be produced by micro-organisms and has provided new methods for increasing the yields of existing ones. The commercial exploitation of the biochemical diversity of micro-organisms has resulted in the development of the fermentation industry and the techniques of genetic manipulation have given this well-established industry the opportunity to develop new processes and to improve existing ones. The term 'fermentation' is derived from the Latin verb *fervere*, to boil, which describes the appearance of the action of yeast on extracts of fruit or malted grain during the production of alcoholic beverages. However, 'fermentation' is interpreted differently by microbiologists and biochemists. To a microbiologist the word means any process for the production of a product by the mass culture of micro-organisms. To a biochemist, however, the word means 'an energy-generating process in which organic compounds act as both electron donors and acceptors', that is, an anaerobic process where energy is produced without the participation of oxygen or other inorganic electron acceptors. In this chapter 'fermentation' is used in its broader, microbiological context.

Microbial Growth

The growth of a micro-organism may result in the production of a range of metabolites but to produce a particular metabolite the desired organism must be grown under precise cultural conditions at a particular growth rate. If a micro-organism is introduced into a nutrient medium which supports its growth, the inoculated culture will pass through a number of stages and the system is termed batch culture. Initially, growth does not occur and this period is referred to as the lag phase and may be considered a period of adaptation. Following an interval during which the growth rate of the cells gradually increases, the cells grow at a constant, maximum rate and this period is referred to as the log, or exponential phase, which may be described by the equation:

$$\frac{dx}{dt} = \mu x \qquad (1)$$

where x is the cell concentration (mg cm^{-3})
t is the time of incubation (h) and
μ is the specific growth rate (h^{-1}).

On integration equation 1 gives

$$x_t = x_o e^{\mu t} \qquad (2)$$

where x_o is the cell concentration at time zero and
x_t is the cell concentration after a time interval, t hours.

Thus, a plot of the natural logarithm of the cell concentration against time gives a straight line, the slope of which equals the specific growth rate which is the maximum for the prevailing conditions and is thus described as the maximum specific growth rate, or μ_{max}. Equations 1 and 2 ignore the facts that growth results in the depletion of nutrients and the accumulation of toxic by-products and thus predict that growth continues indefinitely. However, in reality, as substrate (nutrient) is exhausted and toxic products accumulate, the growth rate of the cells deviates from the maximum and eventually growth ceases and the culture enters the stationary phase. After a further period of time, the culture enters the death phase and the number of viable cells declines. This classic description of microbial growth is illustrated in Figure 1 and it should be remembered that this description refers to the behaviour of both unicellular and mycelial (filamentous) micro-organisms in batch culture, the growth of the latter resulting in the exponential addition of viable biomass to the mycelial body rather than the production of separate, discrete unicells.

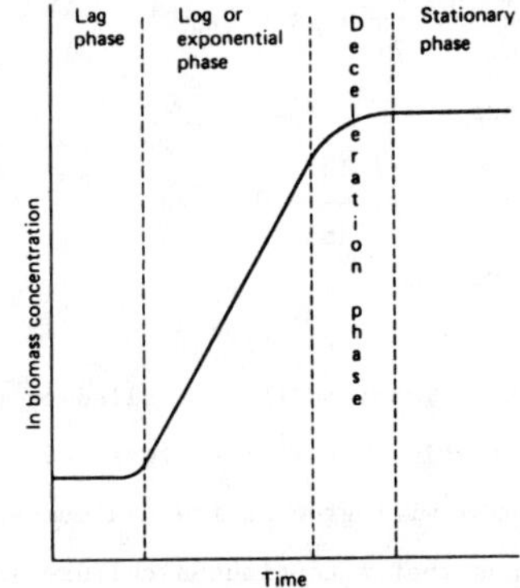

Figure 1. Growth of a 'typical' micro-organism under batch culture conditions (Reproduced with permission from Pergamon Press, Stanbury, P.F. and Whitaker, A., 1984, Principles of Fermentation Technology)

As already stated, the cessation of growth in a batch culture may be due to the exhaustion of a nutrient component or the accumulation of a toxic product. However, provided that the growth medium is designed such that growth is limited by the availability of a medium component, growth may be extended by addition of an aliquot of fresh medium to the vessel. If the fresh medium is added continuously, at an appropriate rate, and the culture vessel is fitted with an overflow device, such that culture is displaced by the incoming fresh medium, a continuous culture may be established. The growth of the cells in a continuous culture of this type is controlled by the availability of the growth-limiting chemical component of the medium and, thus, the system is described as a chemostat. In this system a steady state is eventually achieved and the loss of biomass via the overflow is replaced by cell growth. The flow of medium through the system is described by the term dilution rate, D, which is equal to the rate of addition of medium divided by the working volume of the culture vessel. The balance between growth of cells and their loss from the system may be described as

$$\frac{dx}{dt} = \text{growth} - \text{output}$$

or

$$\frac{dx}{dt} = \mu x - Dx$$

under steady state conditions,

$$\frac{dx}{dt} = 0$$

therefore, $\mu x = Dx$ and $\mu = D$.

Hence, the growth rate of the organisms is controlled by the dilution rate which is an experimental variable. It will be recalled that under batch culture conditions an organism will grow at its maximum specific growth rate and, therefore, it is obvious that a continuous culture may be operated only at dilution rates below the maximum specific growth rate. Thus, within certain limits, the dilution rate may be used to control the growth rate of a chemostat culture.

The mechanism underlying the controlling effect of the dilution rate is essentially the relationship between μ, specific growth rate, and s, the limiting substrate concentration in the chemostat, demonstrated by Monod[1] in 1942:

$$\mu = \mu_{max} \frac{s}{K_s + s} \qquad (3)$$

where K_s is the utilisation or saturation constant which is numerically equal to the substrate concentration when μ is half μ_{max}.

At steady state $\mu = D$

Therefore,

$$D = \mu_{max} \frac{\bar{s}}{K_s + \bar{s}}$$

where $\bar{s}$ is the steady state concentration of substrate in the chemostat

and
$$\bar{s} = \frac{K_s D}{\mu_{max} - D} \qquad (4)$$

Equation 4 predicts that the substrate concentration is determined by the dilution rate. In effect, this occurs by growth of the cells depleting the substrate to a concentration that supports the growth rate equal to the dilution rate. If substrate is depleted below the level that supports the growth rate dictated by the dilution rate the following sequence of events takes place:

i) The growth rate of the cells will be less than the dilution rate and they will be washed out of the vessel at a rate greater than they are being produced, resulting in a decrease in biomass concentration.

ii) The substrate concentration in the vessel will rise because fewer cells are left in the vessel to consume it.

iii) The increased substrate concentration will result in the cells growing at a rate greater than the dilution rate and biomass concentration will increase.

iv) The steady state will be re-established.

Thus, a chemostat is a nutrient limited self-balancing culture system which may be maintained in a steady state over a wide range of sub-maximum specific growth rates.

Fed-batch culture is a system which may be considered to be intermediate between batch and continuous processes. The term fed-batch is used to describe batch cultures which are fed continuously, or sequentially, with fresh medium without the removal of culture fluid. Thus, the volume of a fed-batch culture increases with time. Pirt[2] (1975) described the kinetics of such a system as follows:

If the growth of an organism were limited by the concentration of one substrate in the medium the biomass at stationary phase, x_{max}, would be described by the equation

$$x_{max} \simeq YS_R$$

where Y is the yield factor and is equal to the mass of cells produced per gram of substrate and

S_R is the initial concentration of the growth limiting substrate.

If fresh medium were to be added to the vessel at a dilution rate less than μ_{max} then virtually all the substrate would be consumed as it entered the system:

$$F\,S_R \simeq \mu \frac{X}{Y}$$

where F is the flow rate and

X is the total biomass in the vessel, <u>ie</u> the cell concentration multiplied by the culture volume.

Although the total biomass (X) in the vessel increases with time the concentration of cells, x, remains virtually constant, thus $\frac{dx}{dt} = 0$, $\mu = D$. Such a system is then described as quasi-steady state. As time progresses and the volume of culture increases the dilution rate decreases. Thus, the value of D is given by the expression

$$D = \frac{F}{V_o + Ft}$$

where F is the flow rate

V_o is the initial culture volume and

t is time

Monod[1] kinetics predict that as D falls residual substrate should also decrease resulting in an increase in biomass. However, over the range of growth rates operating the increase in biomass should be insignificant. The major difference between the steady state of the chemostat and the quasi-steady state of a fed-batch culture is that in a chemostat, D is constant whereas in a fed-batch system D decreases with time. The dilution rate in a fed-batch system may be kept constant by increasing, exponentially, the flow rate using a computer control system.

Fermentation Processes

Stanbury and Whitaker[3] classified microbial fermentations into four major groups

i) Those that produce microbial cells (biomass) as the product

ii) Those that produce microbial enzymes

iii) Those that produce microbial metabolites

iv) Those that modify a compound which is added to the fermentation - the transformation processes.

Microbial Biomass. Microbial biomass is produced commercially as single cell protein (SCP) for human food or animal feed and as viable yeast cells to be used in the baking industry. The industrial production of bakers' yeast started in the early 1900's and yeast biomass was used as human food in Germany during the First World War. However, the development of large scale processes for the production of microbial biomass as a source of commercial protein began in earnest in the late 1960's. Several of the processes investigated did not come to fruition due to political and economic problems but the establishment of the ICI pruteen process for the production of bacterial SCP was a milestone in the development of the fermentation industry[4]. This process utilises continuous culture on an enormous scale (1500 m^3) and is an excellent example of the application of good engineering to the design of a microbiological process. The economics of the production of SCP as animal feed are still marginal but ICI are collaborating with Rank Hovis and MacDougalls on a process for the production of fungal biomass to be used as human food and the economics of such a system should prove more attractive.

Microbial Metabolites. The kinetic description of batch culture may be rather misleading when considering the product forming capacity of the culture during the various phases, for, although the metabolism of stationary-phase cells is considerably different from that of logarithmic ones, it is by no means stationary. Bu'Lock *et al.*[5] proposed a descriptive terminology of the behaviour of microbial cells which considered the type of metabolism rather than the kinetics of growth. The term trophophase was suggested to describe the log or exponential phase of a culture during which the sole products of metabolism are either essential to growth such as amino acids, nucleotides, proteins, nucleic acids, lipids, carbohydrates, *etc* or are the by-products of energy yielding metabolism such as ethanol, acetone and butanol. The metabolites produced during the trophophase are referred to as primary metabolites. Some examples of primary metabolites of commercial importance are listed in Table 1.

Table 1. Some examples of microbial primary metabolites and their commercial significance

Primary metabolite	Producing organism	Commercial significance
Ethanol	*Saccharomyces cerevisiae*	'Active ingredient' in alcoholic beverages
Citric acid	*Aspergillus niger*	Various uses in the food industry
Acetone and butanol	*Clostridium acetobutyricum*	Solvents
Glutamic acid	*Corynebacterium glutamicum*	Flavour enhancer
Lysine	*Corynebacterium glutamicum*	Feed additive
Polysaccharides	*Xanthomonas* spp	Applications in the food industry; enhanced oil recovery
Fe^{+++}	*Thiobacillus* and *Sulfolobus*	Ore leaching

The term idiophase was suggested to describe the phase of a culture during which products other than primary metabolites are synthesised, products which do not have an obvious role in cell metabolism. The metabolites produced during the idiophase are referred to as the secondary metabolites. The inter-relationships between primary and secondary metabolism are illustrated in Figure 2, from which it may be seen that secondary metabolites tend to be synthesised from the intermediates and end products of primary metalbolism. Although the primary metabolic routes shown in Figure 2 are common to the vast majority of micro-organisms, each secondary metabolite would be synthesised by very few microbial taxa. Also, not all microbial taxa undergo secondary metabolism; it is a common feature of the filamentous fungi and bacteria and the sporing bacteria but it is not, for example, a feature of the Enterobacteriaceae. Thus, although the taxonomic distribution of secondary metabolism is far more limited than that of primary metabolism, the range of secondary products produced is enormous.

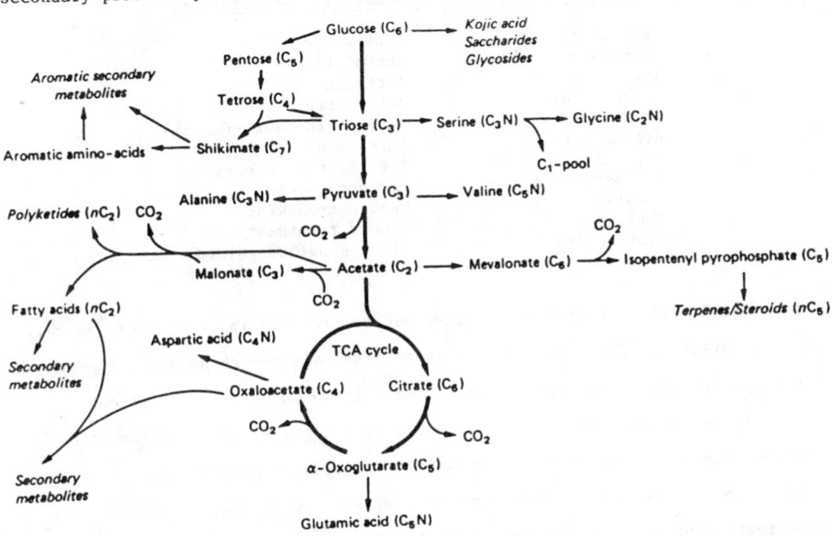

Figure 2. The inter-relationships between primary and secondary metabolism (Reproduced with permission from Academic Press, Turner, W.B., 1971, Fungal Metabolites)

At first sight it may seem anomalous that micro-organisms produce compounds which do not appear to have any metabolic function and are certainly not by-products of catabolism as are, for example, ethanol and acetone. However, many secondary metabolites exhibit antimicrobial properties and, therefore, may be involved in competition in the natural environment[6]; others have, since their discovery in idiophase cultures, been demonstrated to be produced during the trophophase where, it has been claimed, they act in some form of metabolic control[7]. Although the physiological role of secondary metabolism continues to be the subject of considerable debate its relevance to the fermentation industry is the commercial significance of the secondary metabolites. Table 2 summarises some of the industrially important groups of secondary metabolites.

Table 2. Some examples of microbial secondary metabolites and their commercial significance

Secondary metabolite	Commercial significance
Penicillin	Antibiotic
Cephalosporin	Antibiotic
Tetracyclines	Antibiotic
Streptomycin	Antibiotic
Griseofulvin	Antibiotic (anti-fungal)
Actinomycin	Antitumour
Pepstatin	Treatment of ulcers
Cyclosporin A	Immunosuppressant
Krestin	Cancer treatment
Bestatin	Cancer treatment
Gibberellin	Plant growth regulator

The production of microbial metabolites may be achieved in continuous, as well as batch systems. The chronological separation of trophophase and idiophase in batch culture may be studied in continuous culture in terms of dilution rate[8-10]. Secondary metabolism will occur at relatively low dilution rates (growth rates) and, therefore, it should be remembered that secondary metabolism is a property of slow-growing, as well as stationary, cells. The fact that secondary metabolites are produced by slow-growing organisms in continuous culture indicates that primary metabolism is continuing in idiophase-type cells and that secondary metabolism is not switched on to remove an accumulation of metabolites synthesised entirely in a different phase, but that synthesis of the primary metabolic precursors continues through the period of secondary biosynthesis.

The control of the onset of secondary metabolism has been studied extensively in batch culture and, to a lesser extent, in continuous culture. The outcome of this work is that a considerable amount of information is available on the changes occurring in the medium at the onset of secondary metabolism but relatively little is known of the control of the process at the DNA level. Primary metabolic precursors of secondary metabolites have been demonstrated to induce their formation, for example tryptophan in alkaloid[11] biosynthesis and methionine in cephalosporin biosynthesis[12]. On the other hand, medium components have been demonstrated to repress secondary metabolism, the earliest observation being that of Saltero and Johnson[13] in 1953 of the repressing effect of glucose on benzyl penicillin formation. Carbon sources which support high growth rates tend to support poor secondary metabolism and Table 3 cites some examples of this situation. Phosphate and nitrogen sources have also been implicated in the repression of secondary metabolism, as exemplified in Table 3. Therefore, it is essential that repressing nutrients should be avoided in media to be used for the industrial production of secondary metabolites or that the mode of operation of the fermentation maintains the potentially repressing components at sub-repressing levels, as discussed in a later section of this chapter.

Table 3. Some examples of the repression of secondary metabolism by medium components

Medium component	Repressed secondary metabolite
Glucose	Penicillin[13-14]
Glucose	Chloramphenicol[15]
Glucose	Actinomycin[30]
Glucose	Neomycin[31]
Glucose	Streptomycin[32]
Glucose	Cephalosporin[33]
Phosphate	Candicidin[16]
Phosphate	Streptomycin[18]
Phosphate	Tetracycline[65]
Nitrogen source	Penicillin[17]

As mentioned in the introduction, the advent of recombinant DNA technology has extended the potential range of products that may be produced by micro-organisms. Microbial cells may be endowed with the ability to produce compounds normally associated with higher cells and such products may form the basis of new fermentation processes, for example the synthesis of interferon, insulin and renin. It seems to be a widely held opinion that the methods of genetic manipulation will revolutionise the fermentation industry and give rise to a large number of new processes. However, as pointed out by Stanbury and Whitaker[3], the exploitation of these advances depends upon the technology of mass cell culture which has evolved from the yeast and solvent fermentations, *via* the antibiotic fermentations, to the large scale continuous biomass processes. Indeed, in a recent appraisal[19] of the fermentation market in the United States it is claimed that, although the ultimate fruits of genetic engineering are impossible to forecast, the older biotechnology of standard fermentation will remain a highly viable, and predictably growing, method of industrial production. A similar report[20] on the fermentation market in Europe is in press.

Microbial Enzymes. The major commercial utilisation of enzymes is in the food and beverage industries[21] although enzymes do have considerable application in clinical and analytical situations as well as their use in washing powders. Enzymes may be produced from animals and plants as well as microbial sources but the production by microbial fermentation is the most economic and convenient method. Furthermore, it is now possible to engineer microbial cells to produce animal or plant enzymes, for example the production of renin by *E. coli*. Most enzymes are synthesised in the logarithmic phase of batch culture and may, therefore, be considered as primary metabolites. However, some, for example the amylases of *Bacillus stearothermophilus*[22], are produced by idiophase type cultures and may be considered as equivalent to secondary metabolites.

Transformation processes. As well as the use of micro-organisms to produce biomass and microbial products, microbial cells may be used to catalyse the conversion of a compound into a structurally similar, financially more valuable, compound. Although the production of vinegar is the oldest and most well-established transformation process (the conversion of ethanol into acetic acid) the majority of these processes involve the production of high value compounds. The reactions which may be catalysed include oxidation, dehydrogenation, hydroxylation, dehydration and condensation, decarboxylation, deamination, amination and isomerisation[23]. Microbial processes have the advantage of specificity over the use of chemical reagents and of operating under relatively 'mild' conditions. The anomally of the transformation process is that a large biomass has to be produced to catalyse, perhaps, a single reaction. Thus, some processes have been streamlined by immobilising either the cells themselves or the isolated enzymes which catalyse the reactions. This aspect is considered in more detail in Chapter 10.

The Fermentation Process

Figure 3 illustrates the component parts of a generalised fermentation process. Although the central component of the system is obviously the fermenter itself, in which the organism is grown under conditions optimum for product formation, one must not lose sight of operations upstream and downstream of the fermenter. Before the fermentation is started the medium must be formulated and sterilised, the fermenter sterilised and a starter culture must be available in sufficient quantity and in the correct physiological state to inoculate the production fermenter[3]. Downstream of the fermenter the product has to be purified and further processed and the effluents produced by the process have to be treated. Some of these aspects are considered in more detail in Chapter 12.

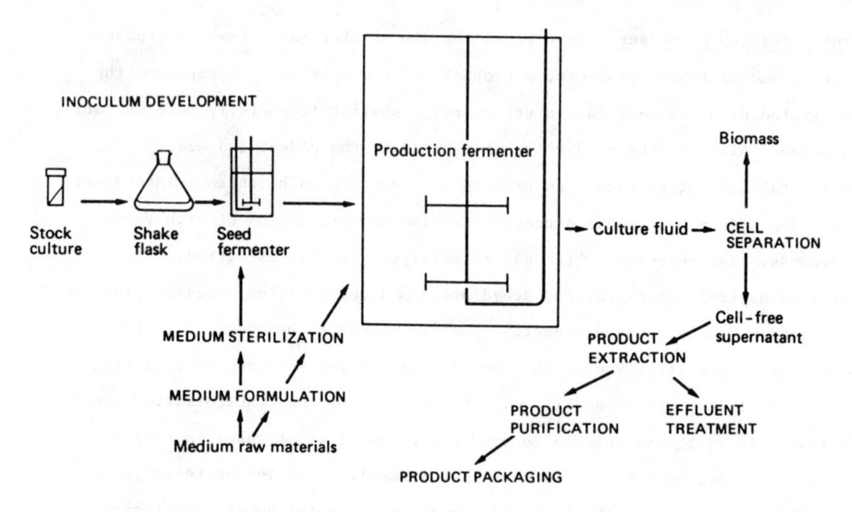

Figure 3. A generalised, schematic representation of a fermentation process (Reproduced with permission from Pergamon Press, Stanbury, P.F. and Whitaker, A. 1984, Principles of Fermentation Technology)

The Mode of Operation of Fermentation Processes. As discussed earlier in this chapter, micro-organisms may be grown in batch, fed-batch and continuous culture and continuous culture offers the most control over the growth of the cells. However, the commercial adoption of continuous culture is confined to the production of biomass and, to a limited extent, the production of potable and industrial alcohol. The superiority of continuous culture for biomass production is overwhelming, as may be seen from the following account, but for other microbial products the disadvantages of the system outweigh the improved process control which the technique offers.

Productivity[3] in a chemostat may be described by the equation:

$$R_{batch} = \frac{x_{max} - x_o}{t_i + t_{ii}} \qquad (5)$$

where R_{batch} is the output of the culture in terms of biomass concentration per hour

x_{max} is the maximum cell concentration achieved at stationary phase

x_o is the initial cell concentration at inoculation

t_i is the time during which the organism grows at μ_{max}

t_{ii} is the time during which the organism is not growing at μ_{max} and includes the lag phase, the deceleration phase and the periods of batching, sterilising and harvesting.

The productivity[3] of a continuous culture may be represented as

$$R_{cont} = D\bar{x} \; 1 - \frac{t_{iii}}{T} \qquad (6)$$

where R_{cont} is the output of the culture in terms of cell concentration per hour

t_{iii} is the time period prior to the establishment of a steady state and includes time for vessel preparation, sterilisation and operation in batch culture prior to continuous operation

T is the time period during which steady state conditions prevail.

Maximum output of biomass per unit time (<u>ie</u> productivity) in a chemostat may be achieved by operating at the dilution rate giving the highest value of $D\bar{x}$, this value being referred to as D_{max}. Batch fermentation productivity, as described by equation 5, is an average for the total time of the fermentation. Because $\frac{dx}{dt} = \mu x$ the productivity of the culture increases with time and, thus, the vast majority of the biomass in a batch process is produced near the end of the log phase. In a steady state chemostat, operating at, or near, D_{max}, the productivity remains constant, and maximum, for the whole fermentation. Also, a continuous process may be operated for a very long time so that the non-productive period, t_{iii} in equation 6, may be insignificant. However, the non-productive time element for a batch culture is a very significant period, especially as it would have to be re-established many times during the running time of a comparable continuous process, and, therefore, t_{ii} would be recurrent.

The steady state nature of a continuous process is also advantageous in that the system should be far easier to control than a comparable batch one. During a batch fermentation heat output, acid or alkali production and oxygen consumption will range from very low rates at the start of the fermentation to very high rates during the late logarithmic phase. Thus, the control of the environment of such a system is far more difficult than that of a continuous process where, at steady state, production and consumption rates are constant. Furthermore, a continuous process should result in a more constant labour demand than a comparable batch one.

A frequently quoted disadvantage of continuous systems is their susceptibility to contamination by 'foreign' organisms. The prevention of contamination is essentially a problem of fermenter design, construction and operation and should be overcome by good engineering and microbiological practice. ICI recognised the overwhelming advantages of a continuous biomass process and overcame the problems of contamination by building a secure fermenter capable of very long periods of aseptic operation, as described by Smith (1980)[4].

The production of growth-associated by-products, such as ethanol, should also be more efficient in continuous culture. However, continuous brewing has met with only limited success and the majority of UK breweries have abandoned such systems due to problems of flavour and lack of flexibility[24]. The production of industrial alcohol, on the other hand, should not be limited by the problems encountered by the brewing industry and continuous culture should be the method of choice for such a process. The adoption of continuous culture for the production of biosynthetic (as opposed to catabolic) microbial products has been extremely limited. Although, theoretically, it is possible to optimise a continuous system such that optimum productivity of a metabolite should be achieved the long term stability of such systems is precarious, due to the problem of strain degradation. A consideration of the kinetics of continuous culture reveals that the system is highly selective and will favour the propagation of the best adapted organism in a culture. 'Best adapted' in this context refers to the affinity of the organism for the limiting substrate at the operating dilution rate. A commercial organism is usually highly mutated such that it will produce very high amounts of the desired product. Therefore, in physiological terms, such commercial organisms are extremely inefficient and a revertant strain, producing less of the desired product, may be better adapted to the cultural conditions than the superior producer and will come to dominate the culture. This phenomenon, termed by Calcott (1981)[25] as contamination from within, is the major reason for the lack of use of continuous culture for the production of microbial metabolites.

Although the fermentation industry has been reluctant to adopt continuous culture for the production of microbial metabolites, very considerable progress has been made in the development of fed-batch systems[26-27]. Fed-batch culture may be used to achieve a considerable degree of process control and to extend the productive period of a traditional batch process without the inherent disadvantages of continuous culture described previously. The major advantage of feeding a medium component to a culture, rather than incorporating it entirely in the initial batch is that the nutrient may be maintained at a very low concentration during the fermentation. A low (but constantly replenished) nutrient level may be advantageous in

i) Maintaining conditions in the culture within the aeration capacity of the fermenter.

ii) Removing the repressive effects of medium components such as rapidly used carbon and nitrogen sources and phosphate.

iii) Avoiding the toxic effects of a medium component.

iv) Providing a limiting level of a required nutrient for an auxotrophic strain.

The earliest example of the commercial use of fed-batch culture is the production of bakers' yeast. It was recognised as early as 1915 that an excess of malt in the production medium would result in a high rate of biomass production and an oxygen demand which could not be met by the fermenter[28]. This resulted in the development of anaerobic conditions and the formation of ethanol at the expense of biomass. The solution to this problem was to grow the yeast initially in a weak medium and then add additional medium at a rate less than the organism could use it. It is now appreciated that a high glucose concentration represses respiratory activity and in modern yeast production plants the feed of molasses is under strict control based on the automatic measurement of traces of ethanol in the exhaust gas of the fermenter. As soon as ethanol is detected the feed rate is reduced. Although such systems may result in low growth rates the biomass yield is near the theoretically obtainable[29].

The penicillin fermentation provides a very good example of the use of fed-batch culture for the production of a secondary metabolite[34]. The penicillin process is a 'two-stage' fermentation; an initial growth phase is followed by the production phase or idiophase. During the production phase glucose is fed to the fermentation at a rate which allows a relatively high growth rate (and therefore rapid accumulation of biomass) yet maintains the oxygen demand of the culture within the aeration capacity of the equipment. If the oxygen demand of the biomass were to exceed the aeration capacity of the fermenter anaerobic conditions would result and the carbon source would be used inefficiently. During the production phase the biomass must be maintained at a relatively low growth rate and, thus, the glucose is fed at a low dilution rate. Phenyl acetic acid is a precursor of the penicillin molecule but it is also toxic to the producer organism above a threshold concentration. Thus, the precursor is also fed into the fermentation continuously thereby maintaining its concentration below the inhibitory level.

The Genetic Improvement of Product Formation

Due to their inherent control systems, micro-organisms usually produce commercially important metabolites in very low concentrations and, although the yield may be increased by optimising the cultural conditions, productivity is controlled ultimately by the organism's genome. Thus, to improve the potential productivity, the organism's genome must be modified and this may be achieved in two ways:

i) Mutation
ii) Recombination.

Mutation. Each time a microbial cell divides there is a small probability of an inheritable change occurring. A strain exhibiting such a changed characteristic is termed a mutant and the process giving rise to it, a mutation. The probability of a mutation occurring may be increased by exposing the culture to a mutagenic agent such as UV light, ionising radiation and various chemicals, for example nitrosoguanidine and nitrous acid. Such an exposure usually involves subjecting the population to a mutagen dose which results in the death of the vast majority of the cells. The survivors of the mutagen exposure may then contain some mutants, a very small proportion of which may be improved producers. Thus, it is the task of the industrial geneticist to separate the desirable mutants (the superior producers) from the very many inferior types. This approach is easier for strains producing primary metabolites than it is for those producing secondary metabolites, as may be seen from the following examples.

The synthesis of a primary microbial metabolite (such as an amino acid) is controlled such that it is only produced at a level required by the organism. The control mechanisms involved are the inhibition of enzyme activity and the repression of enzyme synthesis by the end product when it is present in the cell at a sufficient concentration. Thus, these mechanisms are referred to as feedback control. It is obvious that a good 'commercial' mutant should lack the control systems so that 'overproduction' of the end product will result. The isolation of mutants of Corynebacterium glutamicum capable of producing lysine will be used to illustrate the approaches which have been used to remove the control systems.

The control of lysine synthesis in C. glutamicum is illustrated in Figure 4 from which it may be seen that the first enzyme in the pathway, aspartokinase, is inhibited only when both lysine and threonine are synthesised above a threshold level. This type of control is referred to as concerted feedback control. A mutant which could not catalyse the conversion of aspartic semialdehyde to homoserine would be capable of growth only in a homoserine supplemented medium and the organism would be described as a homoserine auxotroph. If such an organism were grown in the presence of very low

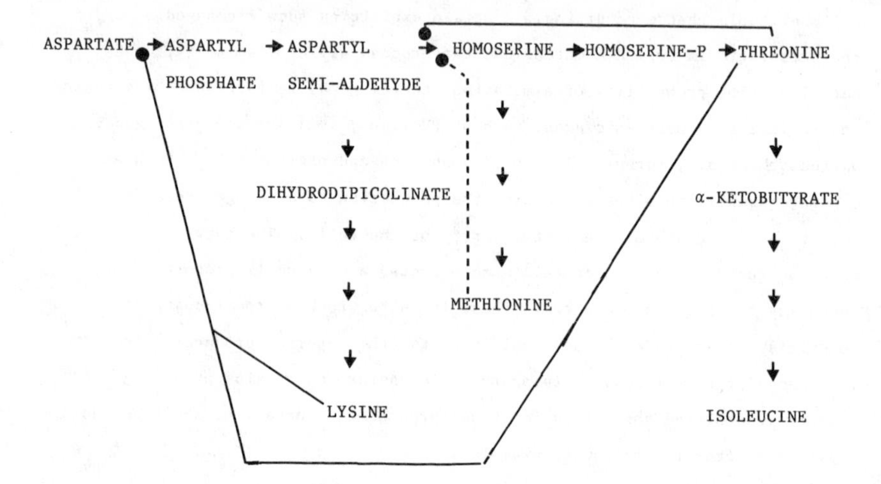

Figure 4. The control of biosynthesis of lysine in Corynebacterium glutamicum

Biosynthetic route →

Feedback inhibition —●

Feedback repression - - -●

concentrations of homoserine the endogenous level of threonine would not reach the inhibitory level for aspartokinase control and, thus, aspartate would be converted to lysine which would accumulate in the medium. Thus, a knowledge of the control of the biosynthetic pathway allows a 'blueprint' of the desirable mutant to be constructed and makes easier the task of designing the procedure to separate the desired type from the other survivors of a mutation treatment.

The isolation of bacterial auxotrophs may be achieved using the penicillin enrichment technique developed by Davis[35]. Under normal culture conditions an auxotroph is at a disadvantage compared with the parental (wild type) cells. However, penicillin only kills growing cells and, therefore, if the survivors of a mutation treament were cultured in a medium containing penicillin and lacking the growth requirement of the desired mutant only those cells unable to grow would survive i.e. the desired auxotrophs. If the cells were removed from the penicillin broth, washed and resuspended in a medium containing the requirement of the desired auxotroph then the resulting culture should be rich in the required type. Nakayama et al.[36] used this technique to isolate a homoserine auxotroph of C. glutamicum which produced 44 g dm^{-3} lysine.

An alternative approach to the isolation of mutants which do not produce controlling end products (ie auxotrophs) is to isolate mutants which do not recognise the presence of controlling compounds. Such mutants may be isolated from the survivors of a mutation treatment by exploiting their capacity to grow in the presence of certain compounds which are inhibitory to the parental types. An analogue is a compound which is similar in structure to another compound and analogues of primary metabolites are frequently inhibitory to microbial cells. The toxicity of the analogue may be due to any of a number of possible mechanisms; for example, the analogue may be incorporated into a macromolecule in place of the natural product resulting in the production of a defective compound, or the analogue may act as a competitive inhibitor of an enzyme for which the natural product is a substrate. Also, the analogue may mimic the control characteristics of the natural product and inhibit product formation despite the fact that the natural product concentration is inadequate to support growth. A mutant which is capable of growing in the presence of an analogue inhibitory to the parent may owe its resistance to any of a number of mechanisms. However, if the toxicity were due to the analogue mimicing the control characteristics of the normal end product, then the resistance may be due to the control system being unable to recognise the analogue as a control factor. Such analogue resistant mutants may also not recognise the natural product and may, therefore, overproduce it. Thus, there is a reasonable probability that mutants resistant to the inhibitory effects of an analogue may overproduce the compound to which the analogue is analagous. Sano and Shiio[37] made use of this approach in attempting to isolate lysine producing mutants of *Brevibacterium flavum*. The control of lysine formation in *B. flavum* is the same as that illustrated in Figure 4 for *C. glutamicum*. Sano and Shiio demonstrated that the lysine analogue S-(2-aminoethyl) cysteine (AEC) only inhibited growth completely in the presence of threonine which suggests that AEC combined with threonine in the concerted inhibition of aspartokinase and deprived the organism of lysine and methionine. Mutants were isolated by plating the survivors of a mutation treatment on to agar plates containing both AEC and threonine. A relatively high proportion of the resulting colonies were lysine overproducers, the best of which produced more than 30 g dm^{-3}. Fuller accounts of the isolation of amino acid and nucleotide producing strains may be found in references 3, 38, 39.

Thus, a knowledge of the control systems may assist in the design of procedures for the isolation of mutants overproducing primary metabolites. The design of procedures for the isolation of mutants overproducing secondary metabolites is more difficult due to the fact that far less information is available on the control of production and, also, that the end products of secondary metabolism are not required for growth. The systems which have evolved, and achieved considerable success, are direct, empirical, screens of the survivors of a mutation treatment for productivity rather than cultural systems which give an advantage to producing types. A typical programme is illustrated in Fig. 5[40].

However, attempts have, and are, being made to adopt a more rational approach to selection techniques and to reduce the empirical nature of the screens. Elander et al.[41] adopted the analogue resistance approach in the isolation of mutants of Pseudomonas aureofaciens overproducing the antibiotic pyrrolnitrin. Tryptophan is a precursor of pyrrolnitrin and a limiting factor of productivity. These workers isolated tryptophan analogue resistant mutants one of which produced 2-3 times more antibiotic than the parental type. Martin et al.[42] removed the inhibitory effect of tryptophan on candicidin production by isolating mutants resistant to tryptophan analogues. Further examples of selection methods for secondary metabolites are given in references 3,43,44,45

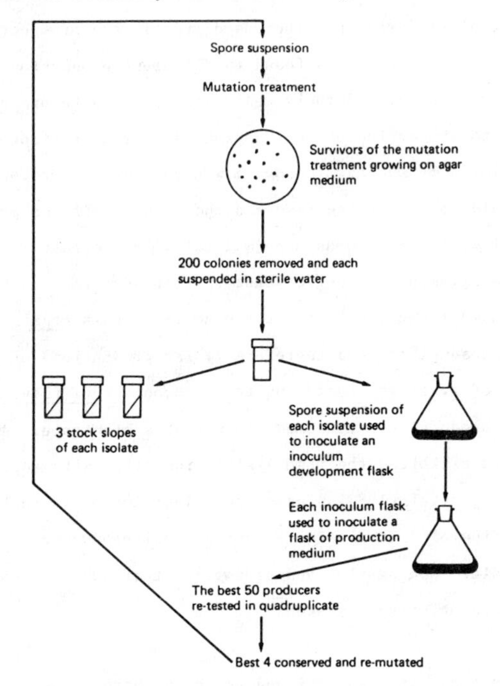

Figure 5. A strain improvement programme for a secondary metabolite producing culture[40]

(Reproduced with permission from Pergamon Press, Stanbury, P.F., and Whitaker, A., Principles of Fermentation Technology)

Recombination. Hopwood (1979)[46] defined recombination as any process which helps to generate new combinations of genes that were originally present in different individuals. Compared with the use of mutation techniques for the improvement of industrial strains the use of recombination has been fairly limited. This is probably due to the success, and relative ease, of mutation techniques and to the lack of basic genetic information on industrial strains. However, techniques are now widely available which allow the use of recombination as a system for strain improvement. In vivo recombination may be achieved in the asexual fungi (for example, Penicillium chrysogenum, used for the commercial production of penicillin) using the parasexual cycle[47]. The technique of protoplast fusion has increased greatly the prospects for combining together characteristics found in different production strains. Protoplasts are cells devoid of their cell walls and may be prepared by subjecting cells to the action of wall degrading enzymes in isotonic solutions. Cell fusion, followed by nuclear fusion, may occur between protoplasts of strains which would not otherwise fuse and the resulting fused protoplast may regenerate a cell wall and grow as a normal cell. Protoplast fusion has been achieved with the filamentous fungi, yeasts, streptomycetes and bacteria and is an increasingly used technique[48-51]. For example, Tosaka et al.[52] improved the rate of glucose consumption (and therefore lysine production) of a high lysine producing strain of B. flavum by fusing it with another B. flavum strain which was a non-lysine producer but consumed glucose at a high rate. Among the fusants one strain exhibited the high lysine production with rapid glucose utilisation. Chang et al. (1982)[53] used protoplast fusion to combine the desirable properties of 2 strains of P. chrysogenum producing penicillin V into one producer strain. Conjugation and protoplast fusion have been used to obtain recombinants of streptomycetes.

In vitro recombination has been achieved by the techniques of in vitro recombinant DNA technology discussed elsewhere in this book. Although the most well-publicised recombinants achieved by these techniques are those bacteria and yeasts which synthesise foreign products, very considerable achievements have been made in the improvement of strains producing conventional products.

The efficiency of Methylomonas methylotrophus, the organism used in the ICI biomass process, was improved by the incorporation of a plasmid containing a glutamate dehydrogenase gene from E. coli[54]. The manipulated organism was capable of more efficient ammonia metabolism which resulted in a 5% improvement in carbon conversion. The production of threonine by E. coli has been improved using in vitro recombination techniques. Debabov (1982)[55] incorporated the entire threonine operon of a threonine analogue resistant mutant of E. coli K_{12} into a plasmid which was then introduced back into the bacterium. The plasmid copy number in the cell was approximately 20 and the activity of the threonine operon enzymes was increased 40-50 times. The organism produced 30 g dm^{-3} threonine compared with the 2-3 g dm^{-3} of the non-manipulated strain. Miwa et al. (1981)[56] applied in vitro recombination to the improvement of threonine producer, E. coli βlM4, which was resistant to α-amino-β-hydroxyvaleric acid (a threonine analogue) and auxotrophic for isoleucine, methionine, proline and thiamine and yielded 3-6 g dm^{-3} threonine. The threonine operon from the production strain was inserted in the plasmid, pBR322, and the hybrid plasmid introduced into a threonine auxotroph derived from the producer, βlM4. The optimisation of the cultural conditions of the recombinant strain resulted in a productivity of 65 g dm^{-3} threonine[57].

The application of the techniques of genetic manipulation in the industrially important Corynebacterium glutamicum has been hindered by the availability of a suitable vector. However, 2 patents were filed in 1983 describing the isolation of corynebacterium plasmids which may be useful in cloning[58-59].

The production of commercially important enzymes may be increased by incorporating the chromosomal gene coding for the enzyme into a plasmid which may then be introduced into the original strain (or a different one) and maintained at high copy number. Colson et al. (1981)[60] cloned a Bacillus coagulans gene coding for a thermostable α-amylase into E. coli where it was replicated and maintained at a high copy number with resultant high enzyme production. Penicillin acylase production in an E. coli strain has also been improved by introducing the relevant gene into a plasmid which was then incorporated into the original strain[61].

The application of in vitro recombinant DNA technology to the improvement of secondary metabolite formation is not as advanced as it is in the primary metabolite field. The main reasons for this relative lack of progress is the ignorance of the basic genetics of secondary metabolite production and the difficulties inherent in the study of compounds which are not essential to the producing cells. However, considerable advances have been made in the genetic manipulation of the streptomycetes in the pioneering work of Hopwood and his co-workers. Techniques have been developed using protoplasts as recipients for genetically modified streptomycete plasmids[62], three of which have been used to clone genes involved in antibiotic synthesis. Although the manipulated cultures did not produce higher levels of antibiotic they should assist in the understanding of the structure and control of secondary metabolism genes. Considerable advances should be made in the investigation of secondary metabolism genetics and several workers have speculated how these might be achieved. Malik[63] has suggested that the isolation of messenger RNA synthesised at the onset of secondary metabolism should provide templates for the synthesis of cDNA, and therefore the genes which may be coding for specific enzymes of secondary metabolism. Kurth and Demain[64] have advocated the use of 'shotgun' cloning either in the producer cells to obtain gene amplification or interspecific which may lead to the synthesis of new antibiotics. A long term aim is to transfer the genes for secondary metabolism from the producer organisms to E. coli and, therefore achieve production with a more 'convenient' organism.

Conclusions

Thus, micro-organisms are capable of producing a wide range of products - a range which has been increased by the techniques of in vitro DNA recombination to include mammalian products. Improved productivity may be achieved by the optimisation of cultural conditions and the genetic modification of the producer cells. However, a successful commercial process for the production of a microbial metabolite depends as much upon chemical engineering expertise as it does on that of microbiology and genetics.

References

[1]J. Monod, Reserches sur les Croissances des Cultures Bacteriennes, Herman and Cie, Paris, 1942

[2]S.J. Pirt, Principles of Microbe and Cell Cultivation, Blackwell, Oxford, 1975

[3]P.F. Stanbury and A. Whitaker, Principles of Fermentation Technology, Pergamon Press, Oxford, 1984

[4]S.R.L. Smith, Phil. Trans. Roy. Soc. (London, B), 1980, 290, 341

[5]J.D. Bu'Lock, D. Hamilton, M.A. Hulme, A.J. Powell, D. Shepherd, H.M. Smalley and G.N. Smith, Can. J. Micro., 1965, 11, 765

[6]A.L. Demain, Search, 1980, 11, 148

[7]I.M. Campbell, Advs. Micro. Physiol., 1984, 25, 2

[8]S.J. Pirt, Chem. Ind., 1968, May, 601

[9]S.J. Pirt and D.S. Callow, J. Appl. Bacteriol., 1960, 23, 87

[10]S.J. Pirt and R.C.Righelato, Appl. Microbiol., 1967, 15, 1284

[11]R.P. Elander, J.A. Mabe, R.L. Hamill and M. Gorman, Fol. Microbiol., 1971, 16, 157

[12]K. Komatsu, M. Mizumo and R. Kodaira, J. Antibiot, 28, 881

[13]F.V. Soltero and M.I. Johnson, Appl. Microbiol., 1953, 1, 2

[14]G. Revilla, J.M. Luengo, J.R. Villanueva and J.F. Martin, Advances in Biotechnology (Editors, E. Vezina and K. Singh), Pergamon Press, Toronto, 1981, Vol. 3, p 155

[15]R.C. Vining and D.S. Westlake, Biotechnology of Industrial Antibiotics (Editor, E.J. Van Damme) Marcel Dekker, New York, 1984, p 387

[16]G. Naharro, J.A. Gill, J.R. Villanueva, J.F. Martin, Advances in Biotechnology (Editors, C. Vezina and K. Singh), Pergamon Press, Toronto, 1981, Vol. 3, p 135

[17]S. Sanchez, L. Paniagua, R.C. Mateos, F. Lara and J. More, ibid, p 147

[18]A.L. Demain, Biotechnology of Industrial Antibiotics (Editor, E.J. Vandamme), Marcel Dekker, New York, 1984, p 33

[19]Frost and Sullivan, Fermentation Markets in USA, Frost and Sullivan, Inc., New York, 1984

[20]Frost and Sullivan, Fermentation Markets in Europe, Frost and Sullivan, Inc., London, In Press

[21]J.T.P. Boing, Prescott and Dunne's Industrial Microbiology, (Editor, G. Reed) MacMillan, New York, 4th Edition, 1982, p 634

[22]A.B. Manning and L.L. Campbell, J. Biol. Chem., 1961, 236, 2951

[23]K. Kieslich, Overproduction of Microbial Metabolites, (Editors V. Krumphanzl, B. Sikyta and Z. Vanek), Academic Press, London, 1984, p 345

[24]B.H. Kirsop, Topics in Enzyme and Fermentation Biotechnology, (Editor, A. Wiseman), Ellis Horwood, Chichester, 1982, p 79

[25]P.H. Calcott, Continuous Culture of Cells, CRC Press, Boca Raton, 1981, Vol. 1, p 13

[26]A. Whitaker, Process Biochem., 1980, 15(4), 10

[27]T. Yamane and S. Shimizu, Advs. In Biochem. Eng./Biotech., 1984, 30, 147

[28]G. Reed and H.J. Peppler, Yeast Technology, Avi, Westport, 1973, p 664

[29]A. Fiechter, Advances in Biotechnology, 1, Scientific and Engineering Principles, (Editors, M. Moo-Young, C.W. Robinson and C. Vezina), Pergamon Press, Toronto, 1982, p 261

[30]M. Gallo and E. Katz, J. Bacteriol., 1972, 109, 659

[31]M.K. Majumdar and S.K. Majumdar, Biochem. J., 1972, 122, 397

[32]E. Inamine, D.B. Lago and A.L. Demain, Fermentation Advances (Editor, D. Perlman) Academic Press, N.Y., 1969, p 199

[33]A. Hinnen and J. Nuesch, Antimicrob. Agents Chemother., 1976, 9, 8242

[34]J.M. Hersbach, C.P. Van der Beek and P.W.M. Van Vijek, Biotechnology of Industrial Antibiotics, (Editor E.J. Vandamme), Marcel Dekker, Inc., New York, 1984, p 45

[35]B.D. Davis, Proc. Natn. Acad. Sci. USA., 1949, 35, 1

[36]K. Nakayama, S. Kituda and S. Kinoshita, J. Gen. Appl. Micro., 1961, 7 (1), 41

[37]K. Sano and I. Shiio, J. Gen. Appl. Micro., 1970, 16, 373

[38]S. Kinoshita and K. Nakayama. Primary Products of Metabolism. Economic Microbiology 2, Academic Press, London, 1978, 210

[39]H. Enei and Y. Hirose, Biotechnology and Genetic Engineering Research Reviews, 2, (Editor G.E. Russell) Intercept, Newcastle upon Tyne

[40]O.L. Davies, Biometrics, 1964, 20, 576

[41]R.P. Elander, J.A. Mabe, R.L. Hamill and M. Gorman, Fol. Microbiol., 1971, 16, 157

[42]J.F. Martin, J.A. Gill, G. Naharro, P. Liras and J.R. Villanueva. In Genetics of Industrial Micro-organisms (Editor O.K. Sebek and A.I. Laskin), American Society for Microbiology, Washington, 1979, p 205

[43]J.F. Martin, Antibiotics and Other Secondary Metabolites. Biosynthesis and Production (Editors, R. Hutter, T. Leisinger, J. Nuesch, W. Wehrlin), Academic Press, London, 1978

[44]R.P. Elander, Biotech. Bioeng., 1980, 22, Supplement 1, 49

[45]A.A. Fantini, Methods in Enzymology, 1975, 43, 24

[46]D.A. Hopwood, Genetics of Industrial Micro-organisms (Editors O.K. Sebek and A.J. Laskin) American Society for Microbiology, Washington, 1979, p 1

[47]K.D. MacDonald and G. Holt, Sci. Prog., 1976, 63, 547

[48]L. Ferenczy, F. Kevei and J. Zsolt, Nature (London), 1974, 793

[49]K. Fodor and L. Alfoldi, Proc. Nat. Acad. Sci. USA, 1976, 73, 2147

[50]D.A. Hopwood, H.M. Wright, M.J. Bibb and S.N. Cohen, Nature (London), 1977, 268, 171

[51]M. Sipiczki and L. Ferenczy, Molecular and General Genetics, 1977, 157, 77

[52]O.Tosaka, M. Karasawa, S. Ikeda and H. Yoshii, Abstract of 4th International Symposium on Genetics of Industrial Micro-organisms, 1982, p 61

[53]L.T. Chang, D.T. Terasaka and R.P. Elander, Dev. Ind. Micro., 1982, 23, 21

[54]J.D. Windon, M.J. Worsey, E.M. Pioli, D. Pioli, P.T. Barth, K.T. Atherton, E.C. Dart, D. Byron, K. Powell and P.J. Senior, Nature, 1980, 287, 396

[55]V.G. Debabov, Overproduction of Microbial Products (Editors V. Krumphanzl, B. Sikyta and Z. Vanek), Academic Press, London, 1982, p 345

[56]K. Miwa, S. Nakamori and H. Momose, Abstracts of 13th International Congress of Microbiology (Boston, USA) p 96

[57]E. Shimizu, H. Heima, T. Osumi, T. Tanaka, K. Miwa, J. Kurashige, S. Nakamori and H. Enei, Abstracts of Annual Meeting of Agricultural Chemical Society, 1983, (Sendai, Japan), p 353

[58]Ajinomoto, European Patent Application, 1983, 71023

[59]Kyowa Hakkokogyo, European Patent Application, 1983, 73062

[60]C. Colson, P. Cornelius, C. Digneffe, R. Walon and C. Walon, European Patent Office, 1981, Patent Application No 81 300 5782

[61]E. Stoppok, U. Schomer, A. Segner, H. Mayer and F. Wagner, Poster Presented at 6th International Fermentation Symposium, 1980, London, Ontario, Canada

[62]D.A. Hopwood, M.J. Bibb, C.J. Bruton, K.F. Chater, J.S. Feitelson and G.A. Gil, Trends in Biotechnology, 1983, 1(2), 42

[63]V.S. Malik, Adv. Appl. Micro., 1982, 28, 28

[64]R. Kurth and A.L. Demain, Biotechnology of Industrial Antibiotics (Editor E.J. Vandamme), Marcel Dekker, New York, 1984, p 791

[65]V. Behal, Z. Hostalek and Z. Vanek, Biotechnol. Lett., 1979, 1, 177.

2

An Introduction to Genetic Engineering

By E. B. Gingold

DIVISION OF BIOLOGICAL AND ENVIRONMENTAL SCIENCES, THE HATFIELD POLYTECHNIC, P.O. BOX 109, COLLEGE LANE, HATFIELD, HERTS. AL10 9AB, U.K.

Over the last decade there has been a revolution in biology. While developments initially centered around studies at the molecular level, few areas of pure and applied biology have remained unaffected. The cause of all this excitement is a series of techniques variously referred to as genetic engineering, gene cloning, *in vitro* genetic manipulation or recombinant DNA technology.

There is no one definition of genetic engineering that satisfies everybody. Nonetheless, describing it as 'the introduction of manipulated genetic material into a cell in such a way as to allow it to replicate and be passed onto progeny cells' covers the main points. That is, the *in vitro* (*i.e.* test tube) manipulation of DNA, the reintroduction of that DNA into living cells and the subsequent establishment of the DNA as part of the hereditary material of the cells. In this way DNA from diverse sources can be introduced into bacterial cells as has been achieved with, to quote well known examples, the human insulin gene[1,2], the interferon genes[3] and the message for human growth hormone[4]. And by linking the foreign genes to bacterial control sequences it is possible to arrange expression of the messages encoded and thereby create bacterial strains with the ability to manufacture these products.

It has been long realised that is is possible to add DNA to cells of certain species of bacteria and change their genetic make up[5]. This process, called transformation, helped to clarify the role of DNA as the genetic material, but there was one severe limitation to this approach: the DNA had to come from the same species and integrate into the chromosome to be passed on.

So what if we had, say, the insulin gene and attempted to add it to bacterial cells? There would be little difficulty in getting the DNA into the cells. Once inside, however, it would not replicate. Even if it were not destroyed by the cell's nucleases (which as a linear fragment it probably would be) it would remain as a single copy. When the initial cell had multiplied to give a million progeny, only one would have the insulin gene!

If we want the gene to be able to replicate, it must be attached to a molecule that is capable of replicating in a bacterial cell. Such a molecule must have a bacterial origin of replication, that is the sequence of DNA recognised by the enzymes of the host cell as a point to initiate replication. A bacterial species such as E. coli will never recognise a eukaryotic origin, hence it is necessary to attach the foreign DNA to a molecule carrying an E.coli origin. Such carrier molecules are called vectors.

Gene Cloning - The basic steps

An outline of the steps involved in gene cloning can be seen in Figure 1. The first step is inserting the foreign DNA into the vector. While the nature of vectors used will be covered in more detail in a later section, a few preliminary words are required here. In theory, it would be possible to remove the main chromosome from the bacterial cell, insert the foreign DNA and return it to a live cell. In practice, this is inconceivable. The chromosome is far too large; it could neither be isolated intact, handled in a test tube nor returned to a cell. What is needed is a far smaller molecule that nonetheless has the ability to replicate autonomously (that is, outside the chromosome).

As it happens, two classes of suitable molecules exist. The first, plasmids, are extrachromosomal molecules of DNA found in many bacterial species[6]. Like the chromosome, they are closed circular molecules although of smaller size. Unlike the chromosome, they are not necessary for the general viability of the cell although they may carry genes, such as those for resistance to antibiotics, that can help survival in special conditions. In addition, some classes of plasmid have what is known as relaxed replication, that is they are present in many copies per cell. Such multi-copy plasmids

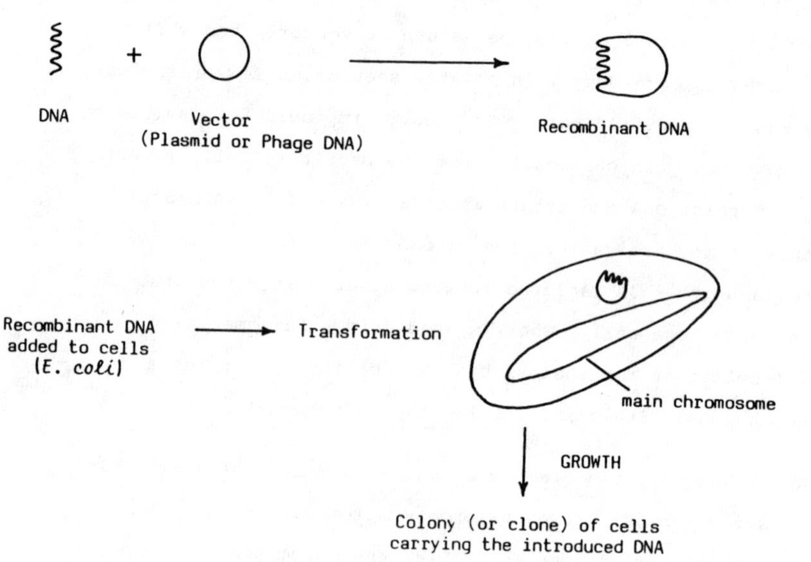

Figure 1 The basic steps in gene cloning

may be conveniently isolated and purified, an essential feature for a cloning vector. The second class of vectors is based on bacterial viruses or phages. Phage DNA can clearly also replicate inside bacteria and give rise to many progeny molecules - this is the basis of the action of phage as an infective agent.

Examining the outline in Figure 1 it is clear that to enable the process to be carried out a number of techniques had to be developed. Essentially these were:

1. A method of reproducibly cutting DNA in the required places.
2. A method of joining (or ligating) DNA fragments.
3. The ability to return the DNA to cells, *i.e.* transformation.
4. A way of selecting for the transformants you require.

The rest of this discussion will examine the 'nuts and bolts' of genetic engineering and show how these problems were overcome. This discussion will concentrate on the use of *E.coli* as the host for manipulated DNA. In later chapters it will become clear how these methods have been extended to a wide range of other species.

Cutting DNA - Restriction Enzymes

At first glance, the ability to cut DNA into smaller fragments presents no difficulty. Anyone who has attempted to isolate DNA will be aware of the fragile nature of the molecule. Pipetting and sonication are just two examples of how DNA can be fragmented, but genetic engineering requires more than this random breakage. It is essential to be able to cut molecules reproducibly and predictably, that is at the same points each time. It was the discovery of a method of doing just this that provided the break-through.

As so often happens in science, the research that opened up this field with its enormous practical applications was, in fact, of the most esoteric kind. The problem under investigation was the ability of bacteria to protect themselves from infection by phage. It was found that bacteria could cut up the phage DNA on entry to the cell provided the phage had previously grown in another bacterial strain[7,8]. Phage DNA previously grown in the same strain was recognised as 'self' and avoided degradation.

It was when the properties of these cutting enzymes were investigated that their practical value was realised. In each case the enzyme recognised a specific sequence, although the sequence differed for enzymes from different species and strains. Furthermore, it was found that the major group of such enzymes (Class II) cut at defined sites within the recognition sequence[9]. Such restriction endonucleases or restriction enzymes are thus an invaluable tool for cutting DNA at defined points.

The recognition sequences for four of the most commonly used enzymes are shown in Figure 2. In fact hundreds of such enzymes have now been identified and an ever increasing number of these are available commercially in a purified form[10]. The four illustrated will, however, serve to demonstrate the general properties. In these examples the recognition sites are either 4 or 6 bases. These are, in general, the most common lengths although 5 and less commonly 7 or more bases may be found in such sequences. As a particular sequence of six bases will occur by chance less frequently than a sequence of four bases, it follows that enzymes that have only four bases in their recognition sequence will cut a given DNA molecule more often than those with six. Thus while EcoRI cuts the phage λ 5 times, Hae III cuts it over 50 times.
Another feature obvious from the figure is that the sequences are symmetrical, that is they have the same series of bases on each strand, though running in opposite directions. This too is a general feature of such enzymes.

The points of cutting are also shown in Figure 2 and from this it should be clear that not all enzymes cut at the centre of the sequence. EcoRI, for example, has staggered cuts and leaves single stranded tails of four bases. Such tails are called cohesive or 'sticky' ends and are of great use in rejoining fragments. The complete range of enzymes includes those that leave 5' and 3' tails of lengths from 2

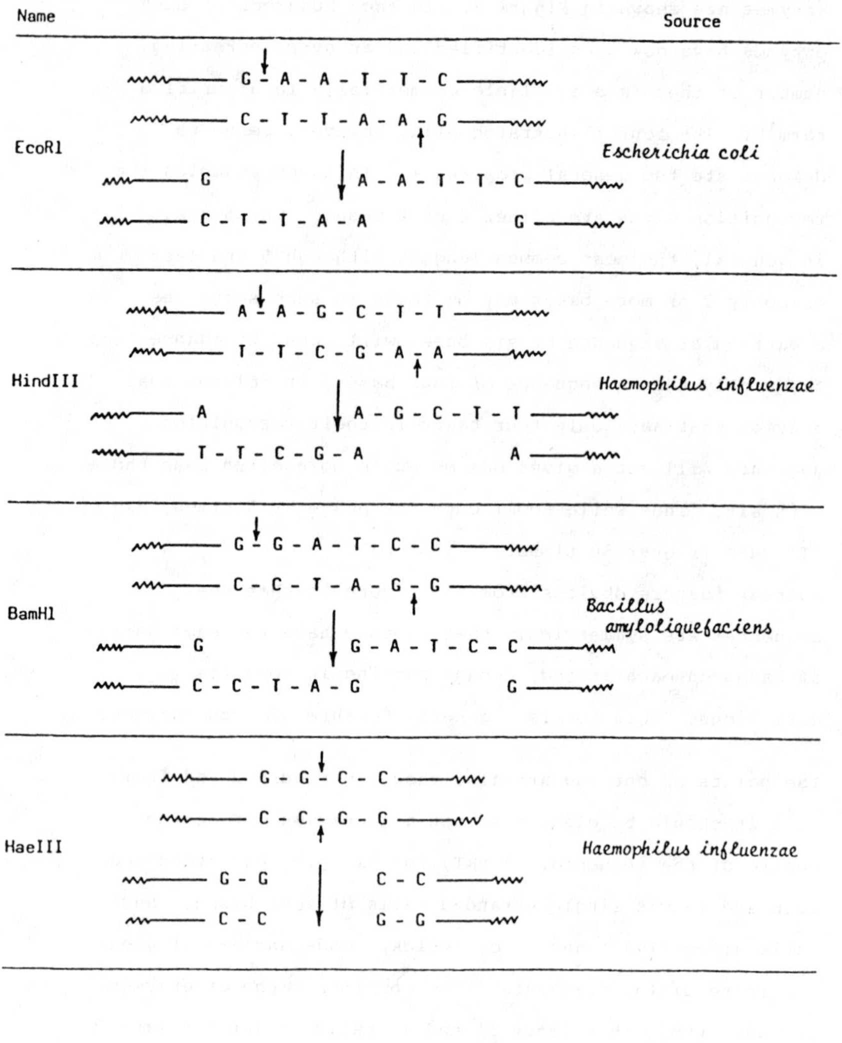

Name		Source
EcoR1	G↓-A-A-T-T-C / C-T-T-A-A↑-G → G A-A-T-T-C / C-T-T-A-A G	*Escherichia coli*
HindIII	A↓-A-G-C-T-T / T-T-C-G-A↑-A → A A-G-C-T-T / T-T-C-G-A A	*Haemophilus influenzae*
BamH1	G↓-G-A-T-C-C / C-C-T-A-G↑-G → G G-A-T-C-C / C-C-T-A-G G	*Bacillus amyloliquefaciens*
HaeIII	G-G↓-C-C / C-C↑-G-G → G-G C-C / C-C G-G	*Haemophilus influenzae*

Figure 2 Some commonly used restriction enzymes

to 5 bases. Other enzymes, such as Hae III, cut centrally and leave blunt ends.

A point of interest is why such enzymes do not attack the DNA of the cells in which they are found. The answer is that a parallel series of enzymes exist that modify the DNA in such cells by adding extra methyl groups to bases within the recognition sequence. Such sequences are thus marked as self and are not subjected to attack. Foreign DNA would, of course, not have such modification, but this problem is avoided in most cloning experiments by using mutant cells lacking in restriction enzymes.

The benefits of being able to cut a DNA molecule into reproducible fragments would be reduced if there were not a convenient method of separating and analysing the fragments. Such separation is generally performed by gel electrophoresis, using agarose gels, or for small fragments polyacrylamide gels[1]. The cut DNA is simply loaded into the wells and allowed to migrate toward the anode. As all molecules should have a similar charge/mass ratio the rate of migration depends simply on the molecular weight. Bands can be visualised with the help of the intercalating agent ethidium bromide under UV light. Using known standards the molecular weights of each fragment can be determined and by a variety of methods including double digestion using two enzymes together, restriction maps (such as those in Figure 3) can be obtained which reveal the positions of each

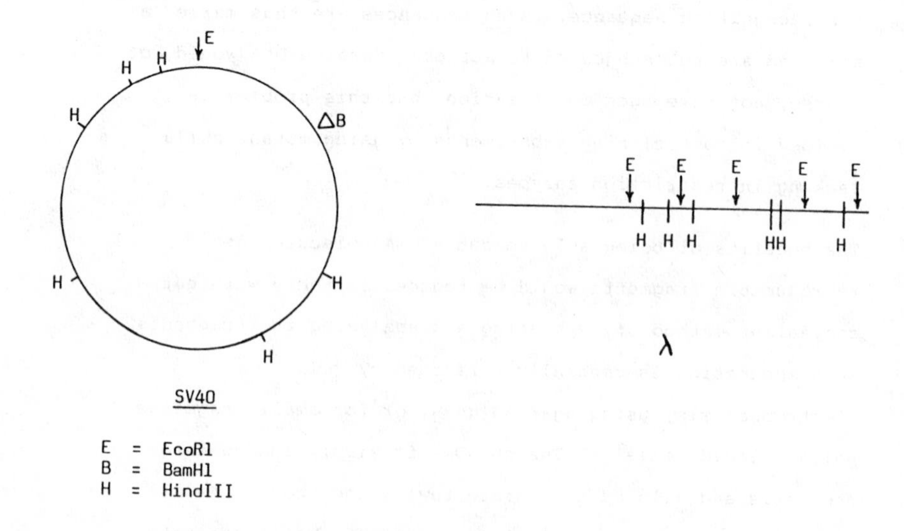

Figure 3 Restriction maps of the phage λ and mammalian virus SV40

fragment on the original molecule and the restriction sites on which the enzymes act.

Joining DNA Molecules

During the early work on gene cloning it was realised that a DNA fragment could be inserted into an open vector if both molecules had complementary single stranded DNA tails. For this reason a poly A tail was artifically added to one molecule and a poly T tail to the other[12]. Later, it was realised that the cohesive ends left by restriction enzymes would also provide suitable complementary regions[13]. The method is illustrated in Figure 4.

With this approach the vector and the DNA to be inserted are cut with the same enzyme (or in some cases different enzymes which nonetheless leave identical single stranded tails). The cut molecules are then mixed under conditions which favour annealing of complementary strands. In fact, the hydrogen bonds involved in 4 sets of base pairs would not be enough to form a stable hybrid. However the enzyme DNA ligase is included in the mix and this will complete the covalent bonds between the molecules in the otherwise transient structure. Thus, as shown in Figure 4, it is possible to form a covalently closed circle including the plasmid and the foreign DNA.

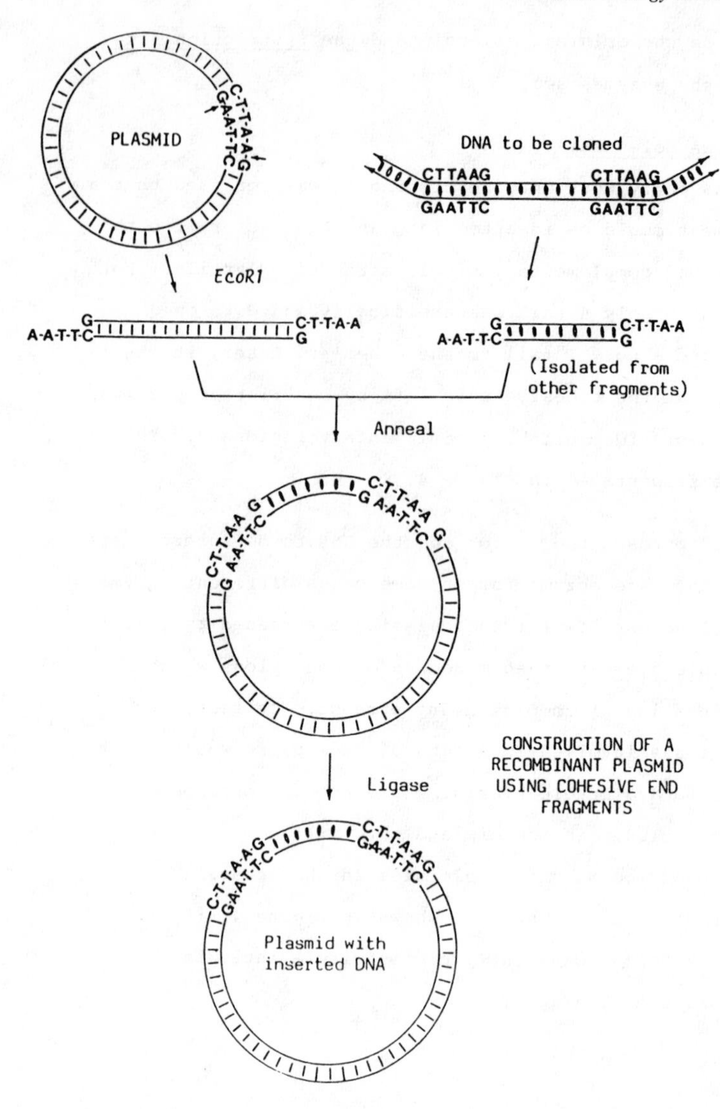

Figure 4 Joining foreign DNA into a plasmid

On further consideration it will become obvious that the desired structure is just one of the possible outcomes of what is a random joining operation. Generally a simple resealing of the vector molecule without inserts is an even more likely outcome. A careful consideration of the kinetics involved allows one to determine concentrations of the molecules to optimalise hybrid formation. In addition, some workers use alkaline phosphatase to remove the phosphate groups from the ends of the single stranded tails of the vector molecules. Without the phosphates, ligase is unable to catalyse the simple recircularisation of these molecules. The foreign DNA to be inserted is not treated with the enzyme and hence retains the ability to be ligated into the vector. As a consequence the vector must incorporate an insert to be successfully recircularised.

There are many situations in which the molecule being inserted into the vector has blunt rather than cohesive ends (that is, no single stranded tails). In fact, the DNA ligase from phage T4 which is generally used in this work is able to join molecules with blunt ends, but only at a low efficiency. This property is used to add small molecules called linkers to the blunt ends as shown in Figure 5. Linkers are synthetic double stranded oligonucleotides that incorporate one or more restriction sites[14]. Despite the low reaction efficiency, the high molar concentration of linker free-ends in the ligation mix ensures that each

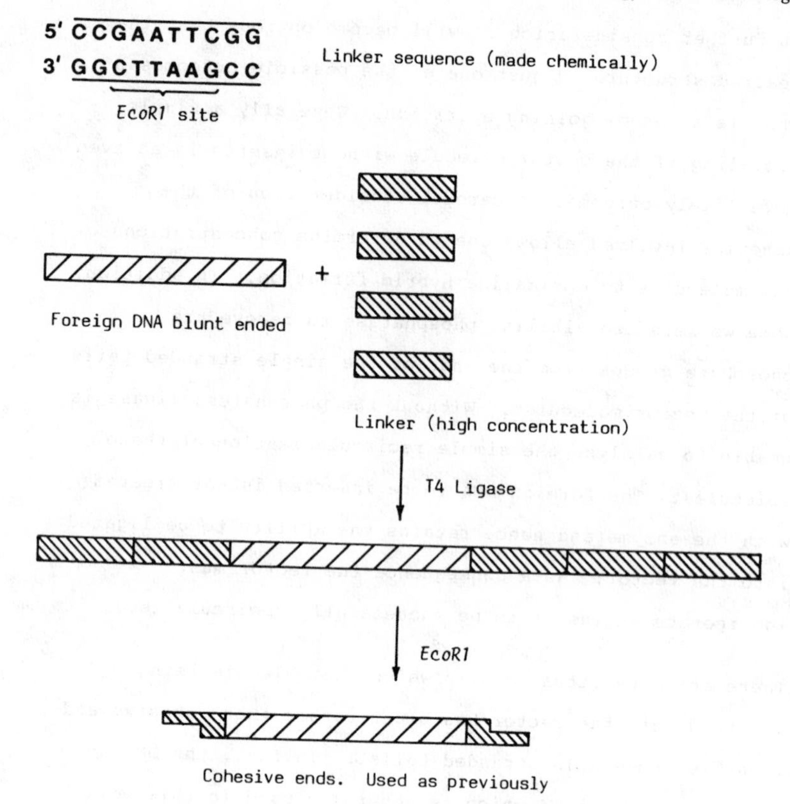

Figure 5 The use of Linkers

molecule of the blunt ended foreign DNA has at least one linker joined to each end. Following treatment with the appropriate restriction enzyme the molecule may now be inserted into an open vector as in the previous method.

Transformation

Once the foreign DNA has been inserted into the vector it becomes necessary to return the DNA into a living cell. For many years it was believed that E.coli could not be transformed by foreign DNA and it is indeed true that this species lacks a natural system for uptake of DNA. Nonetheless, in 1970 it was demonstrated that it is possible to force DNA uptake by a set of extreme conditions[15]. Actively growing cells are harvested and left in hypotonic $CaCl_2$ at 4^oC. After 30 minutes of such treatment changes occur in the cell membrane and the cells are now said to be competent. The DNA is then added to the cell suspension and it is left at 4^oC for a similar period. A short incubation in growth medium then allows cell recovery and plasmid establishment.

A number of improvements have been made to this procedure so as to increase the efficiency of the transformation[16]. Nonetheless, even with the best techniques only a small proportion of cells actually take up the DNA. For this reason it is necessary to have some method of selecting the cells that have been transformed. To this end, all vectors used in this work carry 'selection markers', that is genes which confer easily identifiable phenotypes (characteristics) on cells that have taken them up.

The Nature of Cloning Vectors

1. Plasmids

An ideal plasmid to be used as a cloning vector would, as well as being capable of autonomous replication, be as small as possible so as to enable easy isolation and handling. It would contain single restriction sites (so that it can be opened but not destroyed by the enzymes) and, of course, clear selection markers. An examination of the natural plasmids that were available revealed they were far from this ideal. Most plasmids with useful markers were far too large and contained multiple restriction sites. Some smaller plasmids were known, but these carried genes difficult to use for selection. An early task of the genetic engineers was thus to construct artificial plasmids combining useful features from a number of the natural plasmids in a small molecule.

The plasmid pBR322, shown in Figure 6, has undoubtedly been the most successful of the constructed plasmids[17], and is the 'grand parent' of many of the other vectors used today. An examination of its properties reveals why this is so. The origin of replication incorporated into this plasmid is of the 'relaxed' type, its replication not being under direct control of the main chromosome, and thus multiple

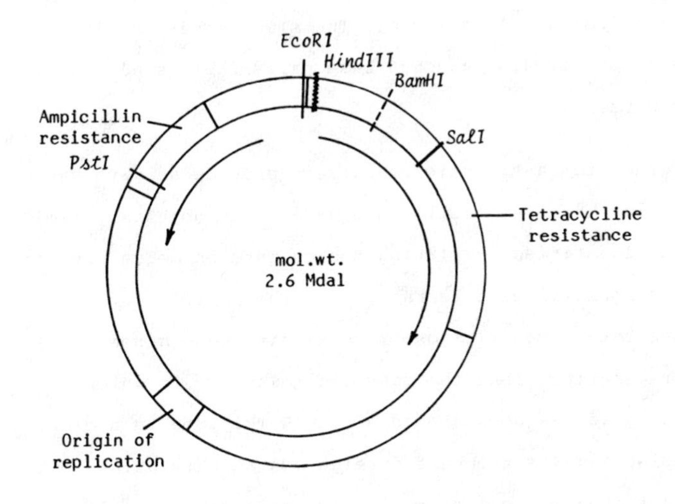

Figure 6 The plasmid pBR322

copies of the plasmid are present per cell. This number can be greatly increased by inhibiting the replication of the main chromosome with the addition of chloramphenicol (so-called amplification). The high copy number is a major factor in the ease of isolation and purification of this plasmid from a cell culture. The small size of 2.6 megadaltons or 4.3 thousand base pairs (kb) is also a major help here.

Two genes for antibiotic resistance provide easy-to-use selection markers. Cells that have taken up this plasmid can be identified by plating the culture on media containing either ampicillin or tetracyline. The position of several of the unique restriction sites within these genes is also a major benefit. Take the case, for example, in which the plasmid has been opened at the BamH1 site in a cloning experiment. Insertion of foreign DNA at this point would disrupt the tetracyline resistance gene while simple recircularisation of the plasmid would not. All transformants would be ampicillin resistant and hence this marker could be used for their selection. The plasmid with inserts, however, would no longer confer tetracycline resistance. Thus a simple test for growth on tetracyline would distinguish the clones carrying inserts (they would not grow) and those that carry the original plasmid. Such a phenomenon is called 'insertional inactivation'.

Derivatives of pBR322 include pAT153, a still smaller vector with an even higher copy number[18]. Other modifications have included the addition of alternative selection markers. Most 'expression vectors', which include the control sequences necessary to enable the product coded for by the foreign DNA to be made, are also pBR322 derivatives.

2. Phage Vectors

While plasmids make good general purpose cloning vectors, there are limitations to their use. The major problem relates to the size of insert which can be efficiently carried into the cell. Transformation frequencies sharply decrease as the size of the DNA circle increases. In addition, plasmids with large inserts are unstable and give rise to smaller segregants. Vectors based on the phage λ have been constructed which can accommodate 15-20 kb inserts and replicate them stably[19]. In this case the cloned product is recovered from plaques, or holes in a bacterial lawn, rather than bacterial colonies. This is because the propagation of the phage lyses the bacterial cells.
A third class of vector, called cosmids, combine properties of plasmids and phages[20]. With these vectors it is possible to add even larger inserts of about 40 kb. Such vectors use the phage infection mechanism to gain entry to the cell but replicate as a large (and unfortunately unstable) plasmid once inside the cell.

The choice of vector is determined by the needs of the particular experiment. For cloning of a single gene, perhaps with the aim of expression, a plasmid might be most convenient. But if the aim was to examine genes in the context of the surrounding sequence on the chromosome, the larger fragments that could be accommodated in a phage vector would determine the choice.

Cloning Actual Genes - The Problem

In this discussion I have so far concentrated on the techniques involved in DNA manipulation. Thus it should now be clear how a piece of foreign DNA can be inserted into a vector and then added to the genetic material of a bacterial cell. What has been avoided is any discussion of how the actual foreign gene is obtained in the first place. This is, in fact, the most difficult part of any cloning experiment. The problem is best illustrated with an example. Suppose you wanted to clone the human growth hormone gene. You could take total human DNA, cut it with an enzyme like EcoRI and insert all resulting fragments into plasmids and then into *E.coli* cells. You would obtain many colonies, each carrying different inserts. Amongst them would be some carrying the human growth hormone gene (or fragments of it if it is cut internally by the enzyme!). But EcoRI cuts human DNA into 700,000 or so fragments. This means that the clones you want would be outnumbered by a factor of 700,000 to 1!

So how do you recognise the clone you are after? Despite the presence of the growth hormone gene it is unlikely to be any larger! In fact, for reasons that should become obvious in later contributions, it is most unlikely that the protein product would be made at all.

It is generally not possible to select the desired clone from the mass of colonies without some kind of a probe. Such probes can be obtained, but a different approach to the problem is needed.

Laboratory Synthesised Genes - Complementary DNA

The genetic material is the same in all cells of the body. Nerve cells, skin cells and liver cells all contain the same genes in the same numbers. What does differ is which genes are active and thus which products are being made.

Developing red blood cells produce great amounts of α and β globin; insulin is produced in the pancrease while it is to the pituitary that one must look for growth hormone production. So while pancrease cells, for example, contain no more copies of the insulin gene than any other cell type, there is a large amount of the insulin mRNA in this tissue. In the case of red blood cells, the enrichment of α and β globin mRNA is so great that it proved possible to isolate these molecules in pure form by classical biochemical techniques.

The mRNA cannot itself be added to a vector and cloned, but a viral enzyme is available which breaks the classical 'central dogma' of genetics and allows us to copy the mRNA message into a complementary DNA sequence[21]. This enzyme, reverse transcriptase, originates from viruses which travel as RNA but convert their genomes into DNA once inside the host cell. The use of the purified enzyme is illustrated in Figure 7.

DNA polymerases require a template, in this case provided by the RNA. They also require a primer, that is, a short length of oligonucleotide already formed at the start of the new molecule. In this reaction, it is possible to use a short oligo-dT molecule for this primer. This will base pair with the poly(A) tail found at the 3' end of eukaryotic mRNA molecules and hence provides a site for initiation of the DNA strand. As well as the reverse transcriptase, the four deoxynucleoside triphosphates must be added to allow synthesis of this complementary DNA strand. The mRNA template can then be hydrolysed away by alkali leaving a single-stranded molecule. The second strand is generally added with DNA polymerase I; as shown in Figure 7, this relies on the original strand forming a hair-pin loop and hence providing a primer. How this looping comes about is not clear, but it probably relies on fortuitous base pairing between regions within the original strand with some degree of homology. The loop itself is then removed with an enzyme

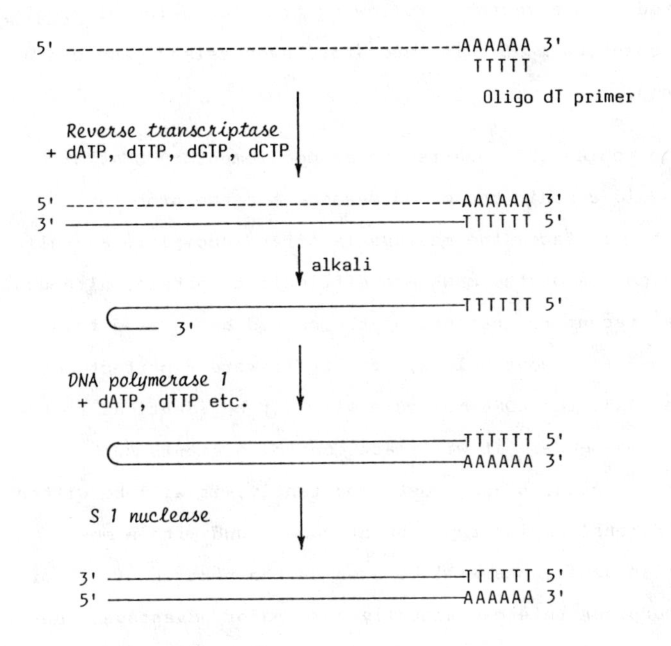

Figure 7 The production of cDNA

such as S1 nuclease that degrades single stranded regions of DNA, cohesive ends added using linkers, and the molecule inserted into a vector. Following transformation of E.coli, large quantities of this laboratory-synthesised gene can be obtained.

Such molecules are referred to as complementary DNA, or cDNA, and should contain the coded message for the protein product. In fact, the message is often incomplete as full length copies of the mRNA are difficult to obtain, although several recent refinements of the method have eased this problem[22,23]. Nonetheless, even if it were a perfect copy of the mRNA, the cDNA molecule will not be identical to the chromosomal gene. It will lack control elements such as promotor sequences and, most importantly, it will be without the 'introns' or interuptions of code found within most mammalian genes. As will be seen in the next chapter, for many purposes this can actually be a major advantage. But for studies on the organisation and control of the real gene it is still necessary to go back and obtain the chromosomal sequence. The cDNA clone gives us the necessary probe for this task.

Colony Hybridisation - Obtaining the Chromosomal Gene

If DNA is heated or treated with alkali it denatures, that is, the strands of the double helix separate. On cooling or returning to neutrality the double helices can reform provided complementary strands come together. The complementary sequences can be from the same original molecule, or different molecules with similar sequences. Provided homology exists, hybrids can form between molecules from diverse sources. The homology between the coding regions of the chromosomal gene and its cDNA copy allows such hybridisation and this forms the basis of this method of detection of the required sequence.

Let us return to the problem of picking out the required clone from the multitude of colonies with DNA from other parts of the genome. In theory, it would be possible to isolate DNA from each colony and test it for hybridisation with a given cDNA preparation. Clearly, however, the numbers involved make this impractical. Using the Grunstein-Hogness method of cloning hybridisation, however, many colonies can be tested at the one time[24]. The colonies are either grown on or, as shown in Figure 8, transferred to a nitrocellulose filter. As the colonies will be destroyed during the experiment it is essential to retain the initial plate or a duplicate. The test filter is then treated with NaOH to lyse the cells and denature the DNA. When the filters are subsequently baked in a vacuum oven the DNA from each colony will become bound to the filter at the position of the original colony.

The cDNA probe is made radioactive by an in vitro process of base replacement called mick translation[25]. This method

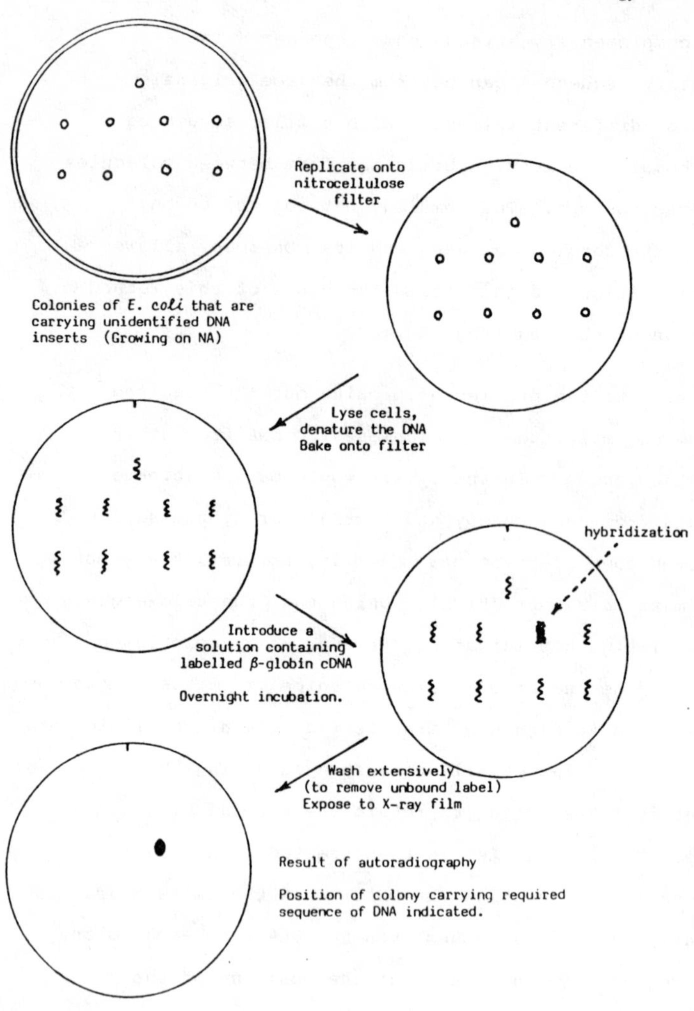

Figure 8 Colony hybridization

allows a highly radioactively labelled molecule to be obtained. It is then denatured by boiling and applied to the filter in an appropriate buffer. An overnight incubation allows the probe to find any complementary sequence and anneal to it. The filter is then extensively washed to remove unbound probe and the positions of bound label revealed by autoradiography. By reference back to the original plate it should now be possible to isolate any colonies with sequences homologous to the cDNA probe and hence carrying the desired chromosomal gene.

With developments in this method it is now possible to screen hundreds of thousands of colonies on a small number of plates[26]. It is thus feasible to screen all the DNA from a given organism cloned into a bacterial host in a search for a particular gene. It has become standard practice to prepare such collections of the total genetic material from a species - so called Genome Libraries - and use them for isolation of individual genes when suitable probes are available. Phage vectors, rather than plasmids, are generally used in the preparation of such libraries as they allow the cloning of larger fragments with greater efficiency and can be stored with greater stability. With such phage systems a technique of plaque rather than colony hybridisation must be used, but the general principles are similar[27].

It can thus be seen that, provided a cDNA clone is available to act as a probe, obtaining a chromosomal gene presents no great difficulty. Unfortunately, however, few cDNA species can be obtained as easily as that for β -globin.

The Probe Problem

In the above discussion of cDNA synthesis the starting material was a purified single mRNA species. There are, however, very few examples of products that are made in such predominance in particular tissues that mRNA purification is possible.

Even in the case of insulin, it was not possible to obtain pure mRNA from pancreatic extracts. More generally, all that is available as the starting material is an mRNA fraction with a small proportion of the desired sequence. When this is used to direct cDNA synthesis and the products are cloned a cDNA library is obtained. But without a pure probe, how do you select the right clones?

A number of approaches have been used to solve this problem. Hybrid released translation (HRT), illustrated as Figure 9, will serve as an example. With this method the DNA from each colony is denatured and bound to a nitrocellulose filter. The total mRNA from the original tissue is then added to the filter under hybridising conditions. Each filter will bind only one mRNA species, the one complementary to the cDNA carried by that particular clone. The remaining mRNA species are then washed off. Using low salt conditions it is possible next to release and collect the bound mRNA. The sample thus collected from each filter is then added to an *in vitro* protein synthesising system and the proteins made in each case are characterised. Gel electrophoresis could be used to determine the molecular weight of the product. Alternatively, immunoprecipitation or an *in vivo* assay of biological activity can be used to provide more certain identification. If a clone carried your required cDNA it would have bound the mRNA for its coded protein. Synthesis of that protein by the later released sample thus reveals the correct clone.

Of course, if the particular clone is rare, the method as outlined above would involve much work. Instead, it is possible to test clones together in 'banks' and then retest the individual members of any 'bank' found to be positive. In this way the cDNA for α interferon was first isolated[3].

More recently the use of chemically synthesised oligonucleotides as probes has become a much used technique. This method requires a knowledge of the amino acid sequence of at least part of the protein product. Using the genetic code a synthetic oligonucleotide corresponding to a run of amino acids is designed and synthesised. This is, however, slightly less straightforward than it might seem due to the degeneracy of the code. As shown in Figure 10

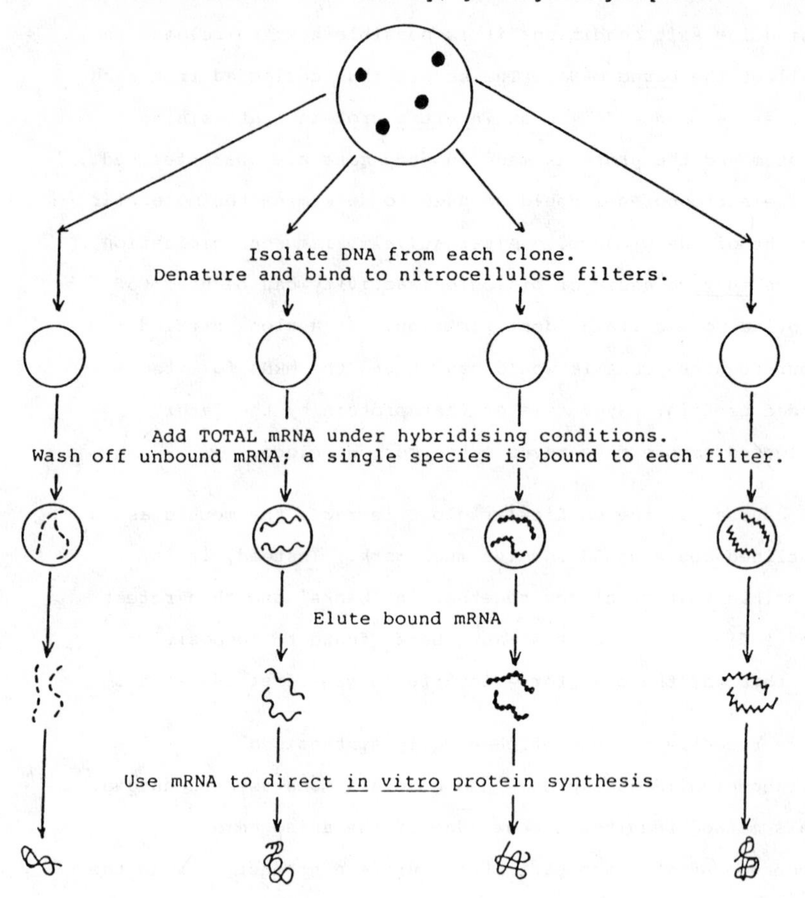

Figure 9 Hybrid Released Translation

```
Amino acid sequence        - Phen -  Trp  -  Pro  -  His  -  Met

                                              T
                                  T           C        T
Nucleotides             5'  T T      T G G  C C    C A      A T C
                                  C           A        C
                                              G
```

Only Trp and Met are unique Codons, therefore 16
possible sequences for the run of amino acids.

Figure 10 Oligonucleotide Primers

there are a number of possible coding sequences for a given stretch of amino acids. While there is no way of knowing which is used in the gene itself before its isolation, the problem can be overcome by using a mixture of nucleotides in the synthesis reactions at the points of degeneracy. In this way the correct complementary sequence will be amongst those synthesised. The probe, once prepared, is used to screen a cDNA library by colony hybridisation and thus pick out the clone with the required message.

With the range of methods available it is possible to consider cloning practically any gene. Obviously the task is easier when the product is relatively abundant and more difficult when it is rare. But when the desire for success is sufficient to overcome the inherent difficulties, it has been reported even for genes producing low level proteins such as the insulin receptor[28].

A Gene in a Clone is Worth.......

In this chapter I have attempted to give an overview of genetic engineering technology, at least in so far as it applies to using *E.coli* as the host. Readers who desire further information are directed to the excellent volumes covering both the theory[29,30,31] and the practice[32,33,34] of genetic engineering. From the rest of this book it should become clear just why this technique has had such dramatic effects on both pure and applied biology.

For the study of basic biology itself, the ability to fish just one gene out of the thousands in a eukaryotic genome, and amplify it by cloning in bacteria, has had profound effects. Once purified, genes can be sequenced by one of two techniques[35,36] and both their coding sequences and control regions studied. Out of such work has come a new understanding of biological phenomena as varied as antibody diversity and the role of oncogenes in cancer.

But for biotechnology itself the main interest is undoubtedly the ability to arrange for foreign messages to be expressed in convenient hosts. It is now possible to conceive of bacteria producing large quantities of substances previously difficult or impossible to obtain in any but minute quantities, e.g. interferon. Other products, already produced and used in large amounts (such as renin), may be obtained more cheaply and conveniently from engineered microorganisms. The next chapter takes up this question of how to switch on the information that has been inserted into the genome of the bacteria.

REFERENCES

1. L. Villa-Komaroff, A. Efstratiadas, S. Broome, P. Lomedico, R. Tizard, S.P. Naber, W.L. Chick and W. Gilbert, Proc. Natl Acad. Sci. U.S.A. 1978, 75 3727.

2. D.V. Goeddel, D.G. Kleid, F. Bolivar, H.L. Heyneker, D.G. Yansura, R. Crea, T. Hirose, A. Kraszewski, K. Itakura and A.D. Riggs, Proc. Natl. Acad. Sci. U.S.A. 1979, 76, 106.

3. S. Nagata, H. Taira, A. Hall, L. Johnsrud, M. Sreuli, J. Ecsodi, W. Boll, K. Cantell and C. Weissmann, Nature, 1980, 284, 316.

4. D.V. Goeddel, H.L. Heyneker, T. Hozumi, R. Arentzen, K. Itakura, D.G. Yansura, M.J. Ross, G. Miozzari, R. Crea and P.H. Seeburg, Nature, 1979, 281, 544.

5. O.T. Avery, C.M. MacLeod and M. MacCarty, J. Exp. Med., 1944, 79, 137.

6. P. Broda, "Plasmids", W.H. Freeman and Company, San Francisco, 1979.

7. D.M. Dussoix and W. Arber, J. Molec. Biol., 1962, 5, 37.

8. S. Lederberg and M. Meselson, J. Molec. Biol., 1964, 8, 623.

9. T.J. Kelly and H.O. Smith, J. Molec. Biol., 1970, 51, 393.

10. R.J. Roberts, Nucleic Acids Res., 1984, 12, 167.

11. E.M. Southern, Methods Enzymol., 1979, 68, 152.

12. D.A. Jackson, R.H. Symons and P. Berg, Proc. Natl. Acad. Sci. U.S.A., 1972, 69, 2904.

13. J.E. Mertz and R.W. Davis, Proc. Natl. Acad. Sci. U.S.A., 1972, 69, 3370.

14. R.H. Scheller, R.E. Dickerson, H.W. Boyer, A.D. Riggs, and K. Itakura, Science, 1977, 196, 177.

15. M. Mandel and A. Higa, J. Molec. Biol., 1970, 53, 159.

16. D. Hanahan, J. Molec. Biol., 1983, 166, 557.

17. F. Bolivar, R.L. Rodriguez, P.J. Greene, M.W. Betlach, H.L. Heynecker, H.W. Boyer, J. H. Crossa and S. Falkow, Gene, 1977, 2, 95.

18. A.J. Twigg and D. Sherratt, Nature, 1980, 283, 216.

19. N.E. Murray, "The Bacteriophage Lambda", Cold Spring Harbour Laboratory, New York, 1983, Vol. 2.

20. J. Collins and B. Hohn, Proc. Natl. Acad. Sci. U.S.A., 1979, 75, 4242.

21. A. Efstratiadis, F.C. Kafotos, A.M. Maxam and T. Maniatis, Cell, 1976, 7, 279.

22. H. Okayama and P. Berg, Molec. Cell Biol., 1982, 2, 161.

23. G. Heidecker and J. Messing, Nucleic Acids Res., 1983, 11, 4891.

24. M. Grunstein and D.S. Hogness, Proc. Natl. Acad. Sci. U.S.A., 1975, 72, 3961.

25. P.W.J. Rigby, M. Dieckmann, C. Rhodes and P. Berg, J. Molec. Biol., 1977, 113, 237.

26. D. Hanahan and M. Meselson, Gene, 1980, 10, 63.

27. W.D. Benton and R.W. Davis, Science, 1977, 196, 180.

28. A. Ullrich, J.R. Bell, E.Y. Chen, R. Herrera, L.M. Petruzzelli, T.J. Dull, A. Gray, L. Coussens, Y.C. Liao, M. Tsubokawa, A. Mason, P.H. Seeburg, C. Grunfeld, O.M. Rosen, and J. Ramachandran, Nature, 1985, 313, 756.

29. R.W. Old and S.B. Primrose, "Principles of Genetic Manipualtion. An Introduction to Genetic Engineering", Blackwell, Oxford, 3rd Edition, 1985.

30. D.M. Glover, "Gene Cloning. The Mechanisms of DNA Manipulation", Chapman and Hall', London, 1984.

31. J.D. Watson, J. Tooze and D.T. Kurtz, "Recombinant DNA. A Short Course", Scientific American books, W.H. Freeman and Company, New York, 1983.

32. T. Maniatis, E.F. Fritsch and J. Sambrook, "Molecular Cloning. A Laboratory Manual". Cold Spring Harbour Laboratory, New York, 1982.

33. J.M. Walker and W. Gaastra, "Techniques in Molecular Biology", Croom-Helm, London, 1983.

34. J.M. Walker, "Methods in Molecular Biology", Humana, New York, Vol. 2, 1985.

35. A.M. Maxam and W. Gilbert, Proc. Natl. Acad. Sci. U.S.A., 1977, 74, 560.

36. F. Sanger, S. Nicklen and A.R. Coulson, Proc. Natl. Acad. Sci. U.S.A., 1977, 74, 5463.

3

The Expression of Foreign DNA in *Escherichia coli*

By R. J. Slater

DIVISION OF BIOLOGICAL AND ENVIRONMENTAL SCIENCES, THE HATFIELD POLYTECHNIC, P.O. BOX 109, COLLEGE LANE, HATFIELD, HERTS. AL10 9AB, U.K.

1. Introduction

The recombinant DNA techniques described in Chapter 2 enable specific DNA sequences to be cloned in bacteria. This means that any DNA sequence, regardless of origin, can be produced in large quantities, thus greatly facilitating studies on gene expression. Restriction endonuclease cleavage sites can be mapped as reference points on the cloned DNA, the nucleotide sequence can be determined and coding sequences (open reading frames, or ORFs) identified. However, unless expression of cloned DNA can be obtained in the host organism, fundamental studies concerned with characterising the function of cloned DNA cannot be carried out, and many of the commercial applications of genetic engineering cannot be achieved.

It has been known since 1944 that it is possible, using clinical techniques of bacterial genetics, to transform bacteria with DNA from the same or closely related species [1]. Recombinant DNA techniques, however, have given molecular biologists the opportunity to obtain expression of foreign DNA from totally unrelated species, even mammals and higher plants, in bacteria such as *Escherichia coli*. The expectation that foreign DNA should be expressed in an unrelated organism is based on the generally held belief that the genetic code is universal. That is, a DNA sequence coding for a protein in one organism should function in any organism to produce a protein with the same amino acid sequence. This belief was given support when it was found that not only DNA from other bacteria but also DNA from the lower eukaryotes *Saccharomyces cerevisiae*[2] and *Neurospora crassa*[3] could transform *E. coli*. Unfortunately, however, when similar experiments were carried out with DNA from higher eukaryotes, expression of the foreign genes was not obtained; more sophisticated techniques, discussed later, were required.

The benefits that accrue from obtaining expression of foreign DNA in *E. coli* are considerable. Firstly, fundamental studies are made possible concerning the relationship between the protein's primary structure (the amino acid sequence) and its function. The DNA sequence of cloned DNA can be altered by *in vitro* mutagenesis and the effect of this mutation on subsequent protein properties studied. The amino acid sequences responsible for catalysis or membrane binding, for example, can then be established. Secondly, the engineering of micro-organisms is possible. The prospect of being able to produce any protein, regardless of origin, in a fermentative organism in culture is a very attractive one, and has obvious commercial implications. In the first instance, proteins used in clinical diagnosis or medical treatment have received the most attention. Insulin production in bacteria is already possible, but this is only the beginning. Interferon, growth hormone, virus coat proteins to form vaccines, produced by micro-organisms, are all likely candidates for early clinical use; but apart from medically important proteins, enzymes such as proteases used by manufacturing and food industries, could be produced on a large scale by genetically engineered bacteria[4]. Chymosin (or rennin), for example, can already be produced in this way for use by the cheese industry [5].

Before the full potential of genetic engineering can be realised, the problems concerned with the expression of DNA from higher eukaryotes in *E. coli* have to be overcome. As was suggested earlier, DNA from an animal or higher plant cannot be used directly to transform *E. coli*. To overcome the problems and obtain expression of eukaryotic DNA in bacteria, it is necessary to understand as much as possible about the control of gene expression in bacteria and higher organisms. This is discussed in the next section.

2. The Control of Gene Expression

The difficulties encountered in obtaining expression of genes from higher eukaryotes in bacteria can be explained if the control systems operating in the different cell types are compared. Although the basic machinery of gene action is the same, that is protein synthesis directed by an RNA copy of a DNA template, there is a significant number of differences between the various types of organism to cause problems for genetic engineers. The mechanisms of gene expression that operate in prokaryotes and eukaryotes are described below and are summarised in Figure 1.

2.1 Prokaryotes. Probably the best understood system of gene control operating in bacteria is the operon model of gene expression originally proposed by Jacob and Monod in 1961 [6]. The model, summarised in Figure 1A, is based on the control of genes responsible for the metabolism of lactose, and is an example of a mechanism of gene expression referred to as "negative control". The lactose, or lac, operon contains three structural genes (i.e. DNA sequences coding for structural proteins or enzymes) referred to as "z", "y" and "a", which code for three enzymes involved in lactose metabolism: β-galactosidase to catalyse the cleavage of lactose to galactose plus glucose, lactose permease that facilitates the entry of lactose into the bacterial cell, and thiogalactoside transacetylase that catalyses the transfer of acetyl groups to galactosides.

The three structural genes are transcribed as a single, so-called polycistronic, mRNA which codes for all three proteins. This is a basic feature of the operon model, giving co-ordinated expression of a number of structural genes, in this case the z, y and a genes. Transcription of the genes is catalysed by DNA dependent RNA polymerase which binds to DNA at the promoter site. Promoters are specific DNA sequences that do not code for proteins but are essential for transcription of structural genes. Promoters are an important factor in the expression of foreign genes in bacteria and are discussed again

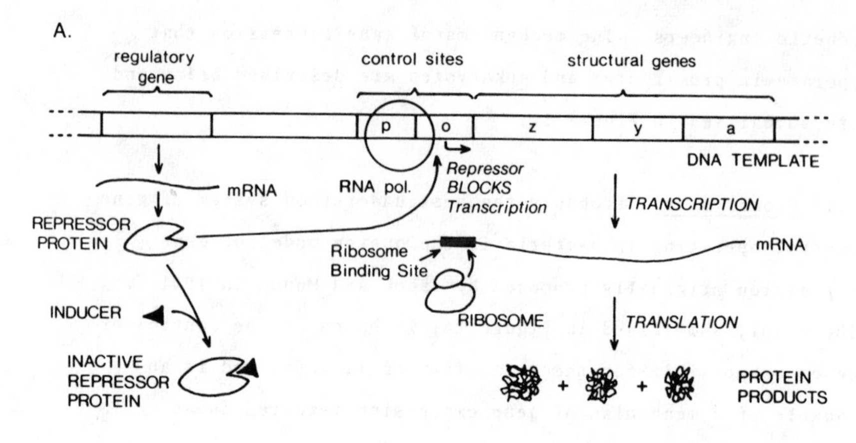

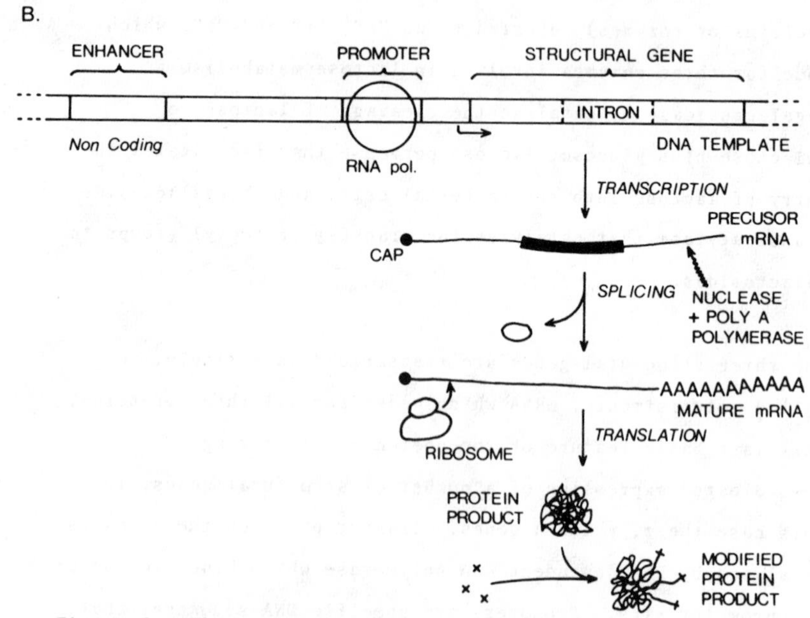

Figure 1. The mechanism of gene expression in prokaryotes (A) and eukaryotes (B).
RNA pol : DNA dependent RNA polymerase
P : promoter
O : operator
↳ : transcription start site
See text for details

later. Transcription begins just "upstream" of the structural genes at a specific initiation point. The RNA transcript contains a nucleotide sequence at the 5' end (called the Shine Dalgarno or S - D sequence) that is complementary to, and can therefore base pair with, RNA in the ribosomes and acts as a ribosome binding site. Translation (protein synthesis) can then proceed. In prokaryotic organisms, protein sysnthesis begins before RNA synthesis is terminated, that is, transcription and translation are "coupled". Downstream of the final structural gene, in this case the "a" gene, there is a termination signal for transcription, usually rich in thymidine nucleotides, where the DNA-RNA polymerase complex is unstable.

Transcription of the structural genes is controlled by a regulatory gene, called the "i" gene in the lac operon, which codes for a protein, referred to as a repressor, that binds to a region of DNA adjacent to the promoter, called the operator. Repressor binding blocks the progress of the RNA polymerase and therefore inhibits transcription, preventing synthesis of enzymes encoded by the structural genes. An inducer acts by binding to the repressor. This alters the repressor's three dimensional structure and prevents it from binding to the operator. In this case the RNA polymerase can continue to transcribe the structural genes coding for lactose metabolism. This system is referred to as negative control because the protein (repressor) that interacts with the operator inhibits transcription.

Negative control is not the only mechanism of gene control acting in bacteria. There are additional control systems such as positive control and a process called attenuation. In positive control, an inducer binds to a protein that stimulates transcription. The best understood example is the system that operates when glucose is not available as an energy source. In this situation, the intracellular concentration of cAMP rises. The cAMP binds to a DNA binding protein called CAP (catabolite activator protein) and the resulting complex binds to the promoters associated with operons involved with the breakdown of alternative sugars. The lac operon is therefore under both negative (lac repressor) and positive (cAMP-CAP) control.

Attenuation refers to incomplete transcription and is a control system dependent on the coupled transcription/translation mechanism that occurs in bacteria. It is a system of control that operates on genes responsible for amino acid synthesis. When a particular amino acid, such as tryptophan, is in short supply, the ribosomes stall at tryptophan codons on the mRNA. This allows particular secondary structures in the mRNA to form that permit transcription to proceed. If the amino acid concerned is in plentiful supply the ribosome does not stall, a different secondary structure forms in the mRNA and transcription of that gene is prematurely terminated.

It is beyond the scope of this article to describe these methods of gene control in any greater depth. Several molecular biology texts [7, 8, 9] are available that give a general overview of the subject and more specific information can be obtained from specialist texts [10] or reviews [11, 12].

It can be seen from the model shown in Figure 1A that many separate elements are required to obtain expression of structural genes: a promoter sequence for RNA polymerase binding, a transcription initiation site, a ribosome binding site and an inducer. The latter can be the naturally occurring inducer, such as allolactose (6-O-β-D-galactopyronosyl -D-glucose, an isomer of lactose in which the galactosyl residue is present on the carbon 6 rather than the carbon 4 of glucose and is produced by basal levels of β-galactosidase) or an artificial inducer such as isopropyl-β-D-thiogalactoside (IPTG). Both of these molecules are active in derepressing the lac operon. The advantage of artificial inducers in stimulating gene expression is that they may show greater activity. In this case, IPTG, active in its native form, is taken up by cells more readily than lactose and is not degraded by β-galactosidase. The lac operon has received so much attention since the operon model for gene expression was first proposed that it has been the obvious choice in the formation of expression vectors, as described later.

2.2 Eukaryotes. The mechanism of gene expression operating in eukaryotes is different in many respects from that in bacteria and is represented in diagramatic form in Figure 1B. It is fair to say that considerably less is known about the control of gene expression in animals and plants in comparison with bacteria and the mechanism of selective gene expression during cell differentiation in multicellular organisms, in which all cells contain the same genetic material, is still a mystery. DNA in higher organisms is maintained in the cell as a complex with histone and other proteins to form a structure referred to as chromatin. Significant alterations in chromatin structure occur during transcription and these changes will be associated with gene control mechanisms. Molecules equivalent to the lac repressor illustrated in Figure 1A have not been discovered in eukaryotes but these are known to be regulatory proteins, such as steroid hormone receptors, which are involved in the control of gene expression by interacting with specific DNA sequences. Although relatively little is known about control mechanisms in eukaryotes, sufficient information is available to explain the problems associated with obtaining expression of eukaryotic DNA in E. coli [13].

No direct equivalent of the operon model has been found in animals and plants. Polycistronic mRNAs do not appear to exist; each structural gene is transcribed separately with its own promoter and transcription initiation and termination sites. The RNA polymerases of eukaryotes are more complex than in bacteria and exist in three forms, specific for rRNA, tRNA and mRNA synthesis. The promoter elements, while serving the same function as in bacteria, contain a different DNA sequence. There does not appear to be a direct equivalent of the operator and once RNA polymerase has bound, RNA synthesis can begin. RNA polymerase binding is influenced by "enhancer" elements which are non-coding, histone-free DNA sequences thought to be responsible for attracting RNA polymerase to coding regions. Following the onset of transcription, the RNA molecule is capped. That is, a 7-methyl guanosine residue is attached by a 5' - 5' phosphate linkage to the end of the RNA.

Transcription is terminated and then significant RNA processing occurs prior to translation; coupled transcription-translation does not appear to occur in eukaryotes.

RNA processing involves several steps. Capping has already been mentioned, but there is also trimming of the RNA with ribonuclease, addition of between 20-250 adenine nucleotides (the "poly(A)tail") to the 3' end and, perhaps most significantly from the point of view of this article, the removal of introns. It has been known since the mid-seventies that many eukaryotic genes contain regions of DNA called intervening sequences, or introns, that do not code for an amino acid sequence. The origin and function of introns is a subject of much debate but the crucial feature is that introns must be removed from the mRNA molecule before protein synthesis can occur. This is carried out by splicing enzymes, present in the nucleus, that remove the intervening sequences and precisely ligate the coding sequences, or "exons", back together again. This has to be carried out in a very precise manner to maintain the correct reading frame of triplet codons for protein synthesis (for an explanation of reading frames see Figure 5). The problem from the point of view of genetically engineering bacteria is that prokaryotes do not possess introns and therefore none of the necessary machinery for their removal. It is not surprising, therefore, that a eukaryotic gene, possessing an intron, cannot be expressed in E. coli.

Additional steps involved in the production of a mature protein in eukaryotes are post-translational modifications such as peptide cleavage, addition of prosthetic groups, glycosylation or formation of multisubunit structures. These are specific to a cell type and are unlikely to be carried out by a bacterial cell. This can cause significant problems in the production of complex eukaryotic proteins by genetically engineered organisms and is likely to be a topic of considerable research interest in the future.

This section has attempted to give some background information which is necessary in order to understand the difficulties involved in obtaining expression of foreign genes in bacteria. The remainder of this article will describe how the problems have been overcome and give some examples of the success to date.

3. The Expression of Eurkaryotic Genes in Bacteria

To obtain expression of eukaryotic genes in bacteria such as *E. coli*, the difference in the mechanism of gene expression between the original organism from which the gene was obtained and the host bacterium must be overcome. The differences in gene expression mechanisms of particular importance to this discussion are: the presence of introns in eukaryotic DNA, the difference in promoter sequences present in bacteria, animals and plants, the absence of a ribosome binding site (Shine-Dalgarno sequence) on eukaryotic mRNA, preferential use of specific triplet codons in coding sequences and, in many cases, the requirement for post-translation modification before the polypeptide is fully functional. The methods used to obviate these difficulties are discussed below.

3.1 Introns. It is apparent from the earlier discussion that a native eukaryotic gene cannot be expressed in bacteria when introns are present. There are two ways in which this problem can be overcome. Firstly, double-stranded DNA copies of mRNA molecules, referred to as complementary DNA or cDNA, can be generated by the use of an mRNA template, reverse transcriptase, DNA polymerase and S1 nuclease *in vitro*, followed by gene cloning *in vivo*. The reverse transcriptase is a viral enzyme that produces a single strand of DNA, complementary in nucleotide sequence to the mRNA; DNA polymerase can then be used to synthesise the second stand and S1 nuclease (a single strand specific enzyme) is used to cut an unwanted DNA loop created by the process (see Figure 2, in the chapter by Maunders, Slater and Grierson in this volume).

The double-stranded cDNA molecule will not contain introns and can act as the coding sequence in expression vectors. There are, however, a number of problems with the cDNA approach. Firstly, if the mRNA is only present as a small constituent of a eukaryotic cell's RNA population, purification of the mRNA can be difficult. Secondly, the cDNA sequence synthesised by reverse transcriptase does not always include the 5' end of the gene; random termination of reverse transcription, prior to completion of complementary strand synthesis, frequently occurs. Thirdly, ligation of additional DNA sequences onto the cDNA is often necessary to tailor the gene for the expression vector; for example restriction enzyme linkers may be required.

A second approach which solves the intron problem is to synthesise the gene by chemical means without a template. If the amino acid sequence of the desired protein product is known, it is possible to chemically synthesise a DNA molecule with the necessary sequence [14].

The advantages of this technique over the cDNA approach are considerable: the complete sequence is obtained, the DNA can be tailored to the vector as desired and particular codons, preferred by the organism chosen as a host for the expression vector, can be incorporated into the gene. There is no theoretical limit to the size of DNA that can be synthesised, but in practice large genes are likely to be synthesised as fragments which are subsequently ligated. This system has already, however, been used to synthesise genes for quite large proteins such as the 514 b.p coding sequence for leucocyte Le (α) interferon, a protein of 166 amino acids [15].

3.2 Promoters. Promoters are sequences of DNA that are necessary for transcription. In *E. coli* the RNA polymerase recognises the promoter as the first step in RNA synthesis. A similar system operates for the transcription of mRNA in eukaryotes but analysis of promoter sequences in bacteria and eukaryotes shows that there are some differences. The nucleotide sequence of some characterised promoters is shown in Figure 2. There is a marked similarity between the various promoter sequences found in *E. coli* represented by the consensus sequence given in Figure 2. The promoters for mRNA synthesis in eukaryotes however, although similar in principle, cannot be efficiently recognised by bacterial RNA polymerase. The important sequences for *E. coli* RNA polymerase are the

	"-35 sequence"		"Pribnow box"	
ptrp	TTGACA	--17bp	--TTAACTA	- - transcription
plac uv5	TTTACA	--18bp	--TATAATG	- - transcription
ptac	TTGACA	--16bp	--TATAATG	- - transcription
prokaryotic consensus	TTGACA		TATAAT	
human B globin	CCAAT	--39bp	--CATAAA	- - transcription
eukaryotic consensus	CCAAT		ATA	

Figure 2 DNA sequence of some characterised promoters

TTGACA ("-35 sequence") and TATAAT ("Pribnow box") sequences found 35 and 10 base pairs upstream of the transcription initiation point respectively.

In order to obtain expression of a eukaryotic DNA sequence in E. coli it is necessary to place the coding sequence downstream from a bacterial promoter. The most commonly used promoters are those from the lac or trp operons or a combination of the two called a "tac" promoter. Many vectors constructed using the lac promoter are based on the principle of using the entire lac control region and the first few nucleotides of the "z" gene coding for β-galactosidase. The foreign DNA to be expressed is then inserted downstream of the β-galactosidase N terminal coding sequence giving rise to a fusion gene, discussed in more detail later. One of the effects of this is to maintain the control mechanism that normally operates on the lac operon; thus expression from the promoter is regulated by the lac repressor. Transcription, or derepression, of the gene is then brought about by the addition of an inducer such as IPTG. Similarly, expression vectors based on the trp promoter incorporate control regions normally in operation for the trp operon.

3.3 Ribosome binding site. Efficient translation of mRNA in prokaryotic cells will not occur without a ribosome binding site. An initiation codon, AUG, is required plus a sequence of 3-9 bases lying between 3 and 12 bases upstream (the S-D sequence) which is complementary to the 3' end of the 16S mRNA molecule (see Figure 3) and is involved in the initiation of protein synthesis by allowing a complex to form between the mRNA and the 30S subunit of the ribosome. Not all mRNAs have an identical S-D sequence but a consensus can be identified. The sequence of the ribosome binding site, together with the secondary structure of the mRNA, has an effect on translation efficiency of mRNA and may have a control function *in vivo*. To obtain expression of foreign genes in *E. coli* it is necessary to incorporate a ribosome binding site into the recombinant DNA molecule. Furthermore, the S-D sequence must be located at the optimal distance from the translation start codon. This is most easily accommodated by the construction of a fusion gene discussed below.

met arg ala

5' GAUUCCU**AGGAGGU**UUGACUAUGCGAGCU --- mRNA

3' AUUCCUCCACUAG 16S rRNA

Figure 3. Base pairing between the Shine-Delgarno sequence (in bold) on the mRNA and a complementary region on the 3' end of the 16S rRNA

3.4 Expression of foreign DNA as fusion proteins. The problems associated with procuring a prokaryotic promoter and S-D sequence to obtain expression of eukaryotic DNA in E. coli can be obviated by constructing a fusion gene. The control region and N terminal coding sequence of an E. coli gene is ligated to the coding sequence of interest. When introduced and cloned in E. coli RNA polymerase will recognise the promoter as native and will transcribe the gene. The 5' end of the mRNA is also native and will consequently interact normally with a ribosome to commence protein synthesis. The protein that results will be chimaeric, the N and C terminals being derived from the prokaryotic and eukaryotic genes respectively. There are a number of advantages in taking this approach. Firstly, expression of the foreign DNA should be efficient. Secondly, the foreign gene can be placed under the control of the induction/repression system of the E. coli promoter/operator used. Thirdly, the fusion peptide may be exported to the periplasmic space via the signal sequence (discussed later) and fourthly, the protein should be relatively stable. This last point is important and worthy of some discussion. For reasons not fully understood, foreign proteins in E. coli are recognised as such and are broken down by endogenous proteases. Foreign peptides expressed as fusion proteins, however, appear to be more stable. The N terminal part of the chimaera is recognised as "self" by the cell. In this respect, the length of the N terminal sequence coded for by E. coli DNA is important; the longer it is the more likely the fusion product is to be stable.

The majority of fusion genes that have been created for expression of foreign DNA in E. coli are based on the lac operon, using the N terminal sequence of β-galactosidase to form the fusion peptide. The first example of this approach was to produce somatostatin in E. coli (16) by the approach illustrated in Figure 4. Somatostatin is a peptide hormone of 14 amino acids with the physiological role of inhibiting secretion of growth hormone, glucagon and insulin. The coding sequence for somatostatin was obtained by chemical synthesis, based on knowledge of the somatostatin amino acid sequence. The DNA was constructed in such a way as to incorporate codons which are preferentially used in E. coli, and to leave single stranded projections, corresponding to the cohesive ends produced by EcoR1 and BamH1 digestion, at each end. The coding sequence was preceded by a codon for methionine and terminated by a pair of nonsense codons to stop translation.

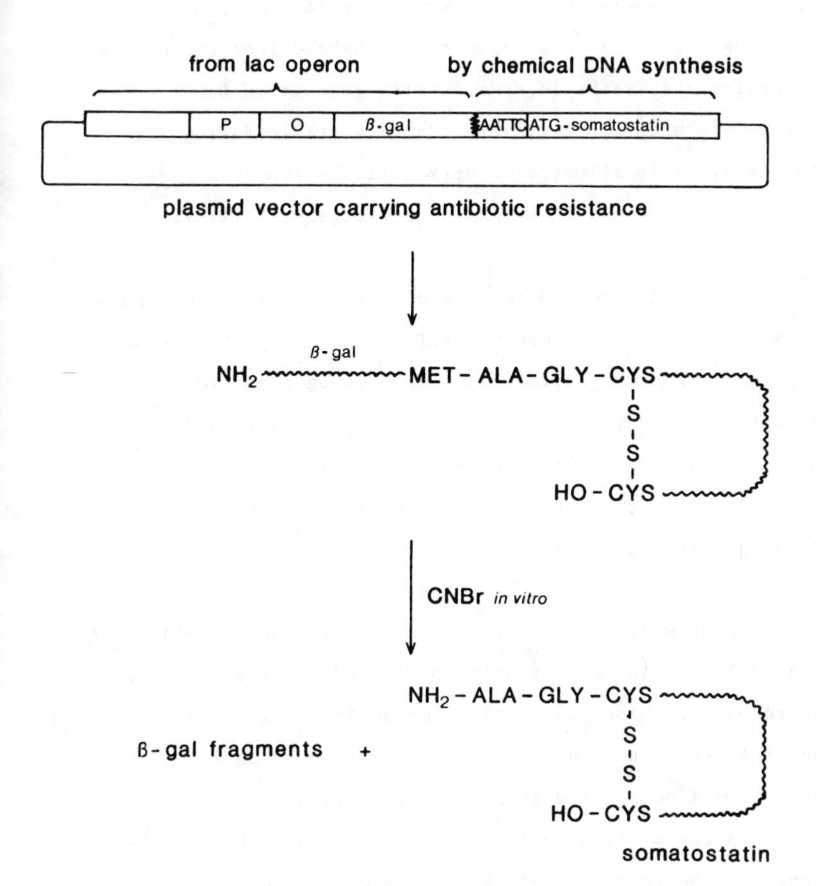

Figure 4. The synthesis in E. coli of somatostatin as a fusion protein.

The initial hybrid genes that were constructed should have produced a hybrid protein containing the first seven amino acids of β-galactosidase, but no somatostatin-like proteins could be detected. When an alternative hybrid gene was created by inserting the synthetic somatostatin gene at an EcoR1 site near the C terminus of β-galactosidase, a stable fusion was synthesised. This illustrates very well the importance of length of fusion genes in maintaining protein stability.

The presence of the methionine residue at the N terminus of the somatostatin amino acid sequence allows for the purification of somatostatin from the fusion protein. Cyanogen bromide treatment in vitro cleaves peptides at the carboxyl side of methionine residues. This process, of course, is only applicable to the production of eukaryotic proteins not containing internal methionines.

A foreign gene will only be expressed as a fusion protein if it is placed in the correct translational reading frame (Figure 5). Since codons are based on triplets of nucleotides there is only a one in three chance that two randomly selected coding sequences will be ligated in phase. To obtain expression, the foreign DNA can be ligated into a position, known by the DNA sequence, to be in the exact reading frame, as in the somotostatin experiment described above, or several recombinant molecules need to be constructed, all in different reading frames, to enable selection of the successful clone. There are different ways of achieving this second approach; for example,

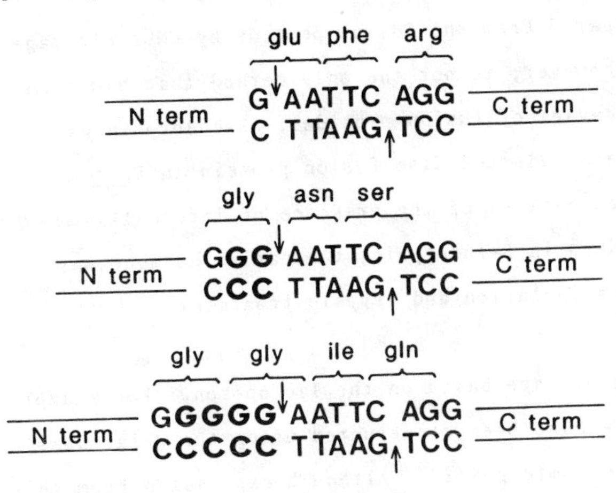

Figure 5. The reading frame, based on triplet codons, of a fusion gene, constructed at an EcoR1 site (arrows), can be altered by the insertion of additional G-C pairs (bold).

successive additions of two G-C base pairs to the EcoR1 fragment of the lac "z" gene has created vectors with three different reading frames [17]. Alternatively, infection of foreign DNA can employ the method of homopolymer tailing, in which the lengths of a linker are randomly constructed [18]. In this case each of the recombinant molecules constructed will have different lengths of the repeating G-C pair and at least some (one in three) should be in the correct reading frame.

The fusion peptide approach has been used in a considerable number of cases following the somatostatin experiment, for example in the synthesis of human insulin [19], thymosin [20], and neo-endorphin [21]. In all these cases the desired peptide was prepared from the fusion peptide by CNBr cleavage. The procedure, however, is not the only method that has been used for the cleavage of fusion peptides. β-endorphin was synthesised as a β-galactosidase fusion protein in E. coli [22]. In this case, because of the presence of internal methionine residues in β-endorphin, the native hormone was prepared by citraconylation and trypsin treatment.

Not all fusion genes are based on the lac operon. For example, fusion genes have also been constructed using the β-lactamase gene carried in plasmid pBR322. Although expression from this promoter is not particularly efficient, its use is of interest because β-lactamase is a secretory protein, responsible for conferring ampicillin resistance, and carries a signal sequence. Signal peptides are sequences of amino acids at the N terminus of proteins; they have an affinity for the cell membrane and are responsible for the export of proteins that carry them. Fusion genes, such as the β-lactamase-proinsulin hybrid, have been constructed at the unique Pst 1 restriction site located between codons 183 and 184 of the β-lactamase coding sequence [18]. The resulting chimaeric proteins contain a substantial proportion of the β-lactamase amino acid sequence and they are, not surprisingly therefore, often found in the periplasmic space. This may have considerable advantages when production of a foreign protein reaches commercial scale. Extraction and purification of the desired proteins should be facilitated and periplasmic proteins may be degraded less than those remaining in the cytoplasm.

3.5 Expression of native proteins. The majority of experiments involving the expression of eukaryotic DNA in bacteria, to date, have employed the construction of fusion genes as a basic strategy. This approach, however, is not mandatory. It is possible to obtain expression of native proteins using a nucleotide sequence that only codes for the peptide required and contains no codons from a prokaryotic gene.

The advantage of this approach in that the amino acid sequence produced should be identical to the naturally occurring eukaryotic protein and therefore should exhibit full biological activity without the need to remove a fusion peptide; a difficult if not impossible task in most cases.

To obtain expression of a native protein, it is necessary to place a coding sequence, with an ATG translation initiation codon, downstream from a bacterial promoter and ribosome binding site. There can be difficulties if a cDNA is being used for the coding sequence. Depending on the nature of the mRNA template used, there may be a long leader sequence upstream of the ATG codon. If there is no convenient restriction enzyme site near the ATG, the leader has to be digested with exonuclease to ensure that a suitable distance between the ATG and ribosome binding site is present in the expression vector. The human growth hormone gene has been successfully expressed in bacteria using the procedure[23] as outlined in Figure 6. The methods used are a good example of how many different techniques or approaches can be combined together in one experiment. The human growth hormone (HGH)

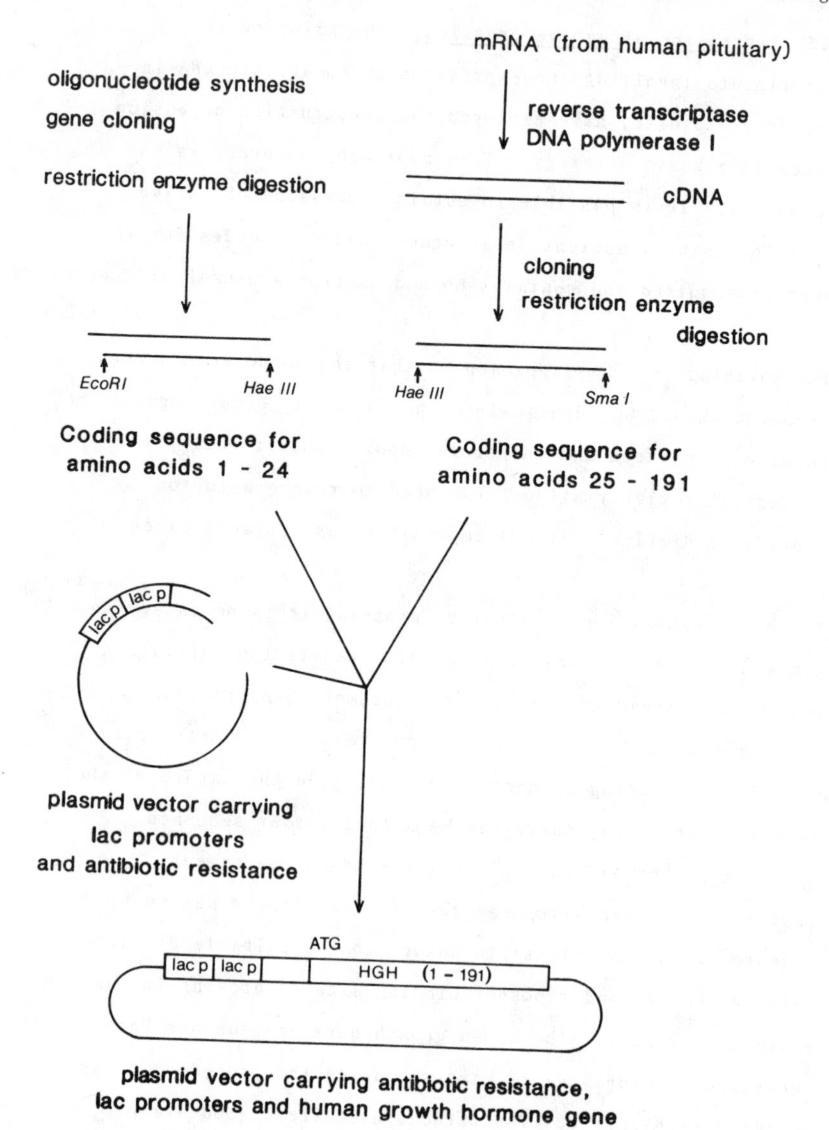

Figure 6. Construction of an expression vector that directs the synthesis of human growth hormone (HGH) in E. coli.

gene was rather too long for complete synthesis by chemical means. The bulk of the coding sequence was therefore obtained by cDNA synthesis and cloning. The cDNA was then cut at convenient restriction enzyme sites to give a defined length, and tailored to the expression vector using a chemically synthesised fragment that coded for the first 24 amino acids of HGH and combined an EcoR1 site for attachment to the plasmid vector. The cDNA and synthetic fragments were ligated together and inserted *via* EcoR1 and Sma1 sites into an expression vector containing two copies of the lac promoter. The resulting plasmid contained a ribosome binding sequence, AGGA, eleven base pairs upstream from the ATG initiaion codon for the HGH gene. In the lac operon, the Shine-Dalgarno sequence lies seven base pairs upstream of the β-galactosidase initiation codon. A derivative of the original expression vector was therefore constructed in which four base pairs between the AGGA and ATG sequences were removed. This was achieved by opening the plasmid with EcoR1 and digesting the single-stranded tails with S1 nuclease and religating the blunt ends. Surprisingly, this new plasmid produced less HGH when introduced into *E. coli* than the original construct, containing the full 11 base pair sequence that had been deliberately shortened. This illustrates the subtleties involved in the relationship between the leader sequence and the initiator codon in protein synthesis and is discussed further in section 5.

The HGH produced in *E. coli* is a soluble protein and can be readily purified. It has the same biological activity as the HGH from human pituitary and apparently differs in only one respect: the presence of an extra methionine residue at the N-terminus. This amino acid would normally be removed by enzymes in the pituitary gland but there is no equivalent activity in E. coli. The presence of this additional methionine does not interfere with the protein's biological activity but, ideally, bacterial products that are likely to be used for clinical purposes need to be as close as possible in structure to their natural counterparts, to avoid complications such as reaction by a patient's immune system.

Human growth hormone is not the only gene that has been expressed as a native protein. Other examples include human leucocyte interferon (LeIF-A), cloned downstream from a trp promoter [24], human fibroblast interferon, using both the trp and lac promoter [25], and mouse dihydrofolate reductase, using a β-lactamase promoter [26]. It is likely that more and more proteins will be produced in this way to simplify the purification of the desired amino acid sequence.

4. Detecting Expression of Foreign Genes

Using the methods of classical microbial genetics, it is relatively simple to detect expression of foreign prokaryotic genes in host organisms. A host is chosen for a transformation experiment that carries a particular mutant, such as a deficiency in the synthesis of histidine. The mutant would therefore be a his^- auxotroph which would normally require a histidine supplement in its growth medium. If the foreign gene to be introduced into the cells codes for histidine synthesis (his$^+$) successful transformants can be detected by their ability to grow without the histidine supplement, i.e. they can be "selected".

This relatively simple technique, however, cannot generally be employed when expression of foreign eukaryotic DNA is desired. The reason is simple: it is unlikely that expression of a eukaryotic gene will confer an advantage to its bacterial host. Insulin synthesis, for example, is hardly likely to enhance the growth advantage of an E. coli cell - unless, of course, the bacterium is diabetic! Alternative methods are required to detect expression of eukaryotic genes. If the function of the desired gene is known a suitable test can be developed. This is relatively straightforward if the the protein required is an enzyme; the usual assay can be applied to a host cell extract. If, however, the desired protein is not an enzyme a more complex test is necessary. These tests usually employ immunodetection techniques [27]. For example, bacterial colonies, suspected of containing the desired protein, are grown on cellulose nitrate paper as a replica of a master agar plate. The colonies are lysed, their contents bound to the cellulose nitrate and the filter incubated with a solution of radioactively labelled antibody, specific to the protein required. Following thorough washing of the filter and autoradiography, colonies containing the required protein can be identified. The relevant colony or colonies on the master plate can then be selected and cultured.

Detecting expression of foreign eukaryotic genes where there is no enzyme assay, or when an antibody is not available, is more difficult. In this case novel proteins over and above the natural background of host proteins need to be detected by protein separation techniques such as polyacrylamide gel electrophoresis. This is rather like looking for a needle in a haystack unless a system is devised to lower the background. Such systems include the use of mini-cells, maxi-cells or a coupled transcription-translation system.

The first two approaches are in vivo methods [28]. Mini-cells are small, spherical cells produced by certain mutant strains of bacteria that carry plasmid DNA, if present in the strain or clone, but no chromosomal DNA. Incorporation of radioactive precursor into RNA or protein in vivo is therefore restricted to a small number of products. Maxi-cells are a strain of E. coli (recA uvrA cells) that react to ultra-violet radiation by degrading chromosomal DNA but leave multi-copy plasmid molecules intact. Therefore, as with mini-cells, expression of plasmid-borne genes can be detected using radioactive precursors in vivo.

The coupled transcription-translation approach employs a soluble in vitro system extracted from E. coli that synthesises proteins from added DNA [29]. In the presence of radioactive amino acids, labelled proteins are produced which can be separated and detected using polyacrylamide gel electrophoresis and autoradiography. Expression vectors can therefore be tested directly for their ability to code for desired proteins. This method is useful when cloning is desired in organisms other than E. coli in which the mini- or maxi-cell techniques are not available.

5. Maximising Expression of Foreign DNA

Until now this article has been concerned with describing the principles and techniques involved in obtaining detectable levels of expression of foreign genes in bacteria. Commercial applications of genetic engineering, however, depend on obtaining high levels of expression such that production of, for example, hormones or vaccines is a realistic economic proposition. The number of cases where genuinely high levels of expression have been obtained is relatively small. Not surprisingly, the achievement of high production levels is a result of considerable research effort and investment. The best examples to date are the production of insulin, growth hormone and interferon with levels of expression approaching 10^5-10^6 molecules per cell [19]. In the case of insulin this is equivalent to nearly 40 mg of product per 100 g wet weight of cells.

Several factors can be identified as being involved in influencing the level of expression. These include: promoter strength, codon usage, secondary structure of mRNA in relation to position of a ribosome binding site, efficiency of transcription termination, plasmid copy number and stability, and the host cell physiology. These factors are discussed below.

Optimal promoters need not necessarily be the naturally occurring ones. Sequences nearest to the consensus sequence are the most efficient, as illustrated by experiments employing the tac promoter [30] a hybrid between the lac and trp promoters (Figure 2). Mutations can be made in promoters to alter their characteristics and thereby influence expression. For example, the L8 and uv5 mutations in the lac promoter render the promoter insensitive to catabolite repression and improve RNA polymerase binding respectively [31].

Codon usage influences levels of expression and can be accommodated if genes are synthesised chemically. The genetic code is degenerate, so for many amino acids there is more than one codon. Efficiency is related to the abundance of tRNA in the cell and the codon-anticodon interaction energy. For a more detailed discussion of codon usage see reviews by Gouy and Gautier [32] and Grosjean and Fiess [33].

The distances between the promoter, ribosome binding site and ATG initiation codon can have a profound effect on levels of expression; greater than 2,000-fold differences have been recorded. This is probably due to different secondary structures forming in the mRNA following transcription. It appears that to obtain high levels of expression the initiation codon (AUG) and the ribosome binding site need to be present as single-stranded structures. The secondary structure of mRNA is dependent on many factors and is difficult to predict. The best approach to the problem, therefore, is to construct a series of vectors with different distances between the S-D and

ATG sequences. A series of clones can then be screened for levels of expression and the optimum selected.

Screening can be difficult if there is no convenient assay for the desired product. A novel solution is to produce a recombinant DNA molecule containing the desired sequence upstream from the carboxy terminal sequence of β-galactosidase placed in the same reading frame. In this case production of a fusion protein can be detected by its β-galactosidase activity. The most suitable of several clones, incorporating varying distances between the S-D and ATG sequences, can then be selected and the carboxy terminal of β-galactosidase removed to leave the required gene sequence as desired [34].

Levels of expression are also influenced by transcription terminator (A-T rich) sequences which must be included at the end of coding regions. Read-through of the RNA polymerase may interfere with other genes downstream or may produce unnecessarily long mRNA molecules that could have reduced translation efficiency and be an undue strain on the cell's energy resources.

Plasmid stability is an important factor in successfully exploiting genetic engineering. There is little to be gained from constructing expression vectors that are lost during large-scale fermentation. Plasmid stability can be maintained by antibiotic selection but this may be undesirable during mass production because of costs and waste disposal problems. The plasmid pBR322 segregates randomly during cell division but naturally occurring plasmids contain a partitioning function, par, which ensures segregation at cell division. Incorporation of par regions into expression vectors may prove to be advantageous to large-scale production [35]. Alternatively, genes could be cloned in expression vectors that confer an advantage to cells carrying the plasmid. Such a gene, for example, might confer resistance to cell lysis by bacteriophages, deliberately included in the growth medium.

Once expression of foreign DNA has been successfully achieved, the host organism will need to be maintained in culture for as long as required. The growth conditions that produce optimum levels of expression at maximum stability for the least expenditure are required. Effects of batch or continuous culture, choice of growth medium *etc.* need to be considered. There has been little systematic work in this area to date.

6. Future Prospects

Clearly there are considerable commercial and scientific opportunities in this area of molecular biology. Many of the basic techniques required to obtain expression of any DNA sequence in any host organism are now available. Future work is likely to concentrate on maximising expression, characterising the best growth conditions for mass production and developing alternative host organisms such as non-pathogenic bacteria like *Bacillus subtilis* that have strains which secrete extracellular proteins thereby facilitating extraction. As discussed earlier, however, bacteria are not capable of carrying out many of the functions required to produce proteins altered by post-translational modifications. In this respect eukaryotic hosts could be exploited. Already there is considerable effort going into the development of yeast as a host organism and for proteins with a high market value there may be a case for cloning in plant or animal cell cultures. Clearly, the future holds much promise and expectation.

7. Further Reading

R.W. Old and S.B. Primrose, "Principles of Gene Manipulation: An Introduction to Genetic Engineering, 3rd Edition". Blackwell, Oxford and Palo Alto. 1985.

D.M. Glover, "Gene Cloning: The Mechanics of DNA Manipulation". Chapman and Hall, London, 1985.

T.J.R. Harris. Expression of eukaryotic genes in E. coli. In "Genetic Engineering 4". Ed. R. Williamson, Academic Press, London and New York, 1983.

8. References

1. O.T. Avery, C.M. Macleod and M. McCarty, J. Exptl. Med., 1944, 79, 137.

2. K. Struhl, J.R. Cameron and R.W. Davis, 1976, Proc.Nat.Acad.Sci. USA, 73, 1471.

3. D. Vapnek, J.A. Hautala, J.W. Jacobson, N.H. Giles and S.R. Kushner, 1977, Proc.Nat.Acad.Sci. USA, 74, 3508.

4. W. Gilbert and L. Villa-Komaroff, Sci.Am., 1980, 242, 74.

5. T. Beppu. Trends in Biotechnology, 1983, 1, 85.

6. F. Jacob and J. Monod, J. Mol. Biol, 1961, 3, 318.

7. D. Friefelder, "Molecular Biology", Science Books International, 1983.

8. B. Lewin, "Genes", Wiley, 1983.

9. J.D. Watson, J. Tooze and D.T. Kurtz, "Recombinant DNA: A Short Course". Scientific American Books, 1983.

10. R.E. Glass, "Gene Function: E. coli and its heritable elements". Croom Helm, London and Canberra, 1982.

11. M. Rosenberg and D. Carst, Ann. Rev. Genet., 1979, 13, 319.

12. I.P. Crawford and G.C. Slauffer, Ann. Rev. Biochem., 1980, 49, 163.

13. J.R. Nevins, Ann. Rev. Biochem., 1983, 52, 441.

14. M.H. Caruthers, S.L. Beaucage, C. Becker, W. Efcavitch, E.F. Fisher, G. Galluppi, R. Coldman, P. Dettaseth, F. Martin, M. Matteucci and Y. Stabinsley, in "Genetic Engineering". Eds. J.K. Setlow and A. Hollaender, Plenum Press, New York and London, 1982, 119.

15. M.D. Edge, A.R. Green, G.R. Heathcliffe, P.A. Meacock, W. Shuch, D.B. Scanlon, T.C. Atkinson, C.R. Newton and A.F. Markham, Nature, 1981, 191, 756.

16 K. Itakura, T. Hirose, R. Crea, A.D. Riggs, H.L. Heyneker, F. Bolivar and H.W. Boyer, Science, 1977, 198, 1056.

17. P. Charney, M. Perricaudet, F. Galibert and P. Tiollais, Nucleic Acid Res., 1978, 5, 4479.

18. L. Villa-Komaroff, A. Efstratiadas, S. Broome, P. Lomedico, R. Tizard, S.P. Naber, W.L. Chick and W. Gilbert, Proc. Nat. Acad. Sci. USA, 1978, 75, 3727.

19. D.V. Goeddel, D.G. Kleid, F. Bolivar, H.C. Heynecker, D.G. Yansura, R. Crea, T. Hirose, A. Kraszeuski, K. Itakura and A.D. Riggs, Proc. Nat. Acad. Sci. USA, 1979, 76, 106.

20. R. Wetzel, H.L. Heyneker, D.V. Goeddel, G.B. Thurman and A.L. Goldstein, Biochemistry, 1980, 19, 6096.

21. S. Tanaka, T. Oshima, K. Ohsue, T. Ono, S. Oikawa, I. Takano, T. Noguchi, K. Kangawa, N. Minamino and H. Matsuo, Nucleic Acids Res., 1982, 10, 1741.

22. J. Shine, I. Fettes, N.C.Y. Lan, J.L. Roberts and J.D. Baxter, Nature, 1980, 285, 456.

23. D.V. Goeddel, H.L. Heyneker, T. Hozumi, R. Arentzen, K. Itakura, D.G. Yansura, M.J. Ross, G. Miozzari, R. Crea and P.H. Seeburg, Nature, 1979, 281, 544.

24. D.V. Goeddel, E. Yelverton, A. Ullrich, H.L. Heyneker, G. Miozzari, W. Holmes, P.H. Seeburg, T. Dull, L. May, N. Stebbing, R. Crea, S. Maeda, N. McCandliss, A. Sloma, J.M. Tabar, M. Cross, P.C. Familletti and S. Pestka, Nature, 1980, 287, 411.

25. T. Taniguchi, L. Guarente, T.M. Roberts, D. Kimelman, J. Douhan and M. Ptashne, Proc. Nat. Acad. Sci. USA, 1980, 77, 5230.

26. A.C.Y. Chang, H.A. Erlich, R.P. Gunsalus, J.H. Nunberg, R.J. Kaufman, R.T. Schimke and S.N. Cohen, Proc. Nat. Acad. Sci. USA, 1980, 77, 1442.

27. S. Broome and W. Gilbert, Proc. Nat. Acad. Sci. USA, 1978, 75, 2746.

28. B. Oudega and F.R. Mooi, in "Techniques in Molecular Biology". Ed. J.M. Walker and W. Gaastra, Croom Helm, London and Canberra, 1983, Chapter 13, p239.

29. J.M. Pratt, in "Transcription and Translation, A Practical Approach". Ed. B.D. Hames and S.J. Higgins, IRL Press, Oxford and Washington D.C. 1984, Chapter 7, p179.

30. H.A. de Boer, L.J. Comstock and M. Vasser, Proc. Nat. Acad. Sci. USA, 1983, 80, 21.

31. F. Fuller, Gene, 1982, 19, 43.

32. M. Gouy and C. Gautier, Nucleic Acid Res, 1982, 10, 7055.

33. J. Grosjean and W. Fiers, Gene, 18, 1982, 199.

34. L. Guarente, G. Lauer, T.M. Roberts and M Ptashne, Cell, 1980, 20, 543.

35. G. Skogman, J. Nilsson and P. Gustafsson, Gene, 1983, 23, 105.

4

Cloning in Brewer's Yeast, *Saccharomyces cerevisiae*

By D. H. Williamson

NATIONAL INSTITUTE FOR MEDICAL RESEARCH, THE RIDGEWAY, MILL HILL, LONDON NW7 1AA, U.K.

Transformation; Procedures for Inducing Uptake of Exogenous DNA

The first, essential, step in cloning DNA in any organism is to induce the cells concerned to take up exogenous DNA molecules. In the case of brewer's yeast, there are two main procedures for achieving this.

In the first of these[1,2], the thick cell wall is partially destroyed by an enzyme and the resulting osmotically-sensitive spheroplasts are exposed to DNA molecules in the presence of polyethylene glycol (PEG), an agent commonly used for fusing cells and protoplasts. If the fused spheroplasts are embedded in agar containing an osmotic stabiliser, a proportion of them are able to regenerate whole cells, and a fraction of these regenerant individuals can be shown to have taken up DNA molecules. Selection of the successful transformants is normally achieved by using a mutant host auxotrophic for a given nutrient, and a vector carrying a gene that complements that mutation - the agar suspension of spheroplasts being overlayed on medium lacking the nutrient in question. Depending on the particular vector and host strain used, up to 20,000 transformants may be recovered per µg of transforming DNA. In the absence of selection, 20 - 40% of the spheroplasts may be capable of regenerating cells, but normally less than 1% of these will be found to be transformants.

An alternative transformation procedure avoids cell wall dissolution and PEG treatment, and requires only that the recipient cells are treated with LiCl[3]. This is generally less efficient (by 1-2 orders of magnitude) but the overall simplicity of the procedure, and the ability to plate the cells without embedding or the use of an osmotic stabiliser, makes it a favourite for many applications.

Detailed recipes can be found in the references cited above. It is however worth pointing out that strains vary widely in their susceptibility to transformation by either of these two methods, and a certain amount of old fashioned genetic analysis - mating and selection of haploid progeny-may be required for construction of a transformable strain of a given genotype.

Vectors, Modes of Transformation, and Applications

The wide range of vectors available for cloning in yeast fall into a few broad categories which have quite different properties and applications. These are summarised in Figure 1.

Of necessity, all the vectors carry at least one gene permitting selection in yeast and, to facilitate bulk production, most of them are hybrid 'shuttle' vectors, having *E.coli* plasmid sequences to allow selection and amplification in that organism. This is desirable since yeast plasmids cannot be highly amplified, and yields are usually poor. Normally, selection in *E.coli* makes use of ampicillin or tetracycline resistance genes on the plasmid, but some yeast genes (notably *LEU2*, *HIS3*, *URA3* and *TRP1*) function in *E.coli* and can be used for selection in that organism. Commonly, inserted DNAs are ligated into a unique restriction site in amp^R or tet^R, insertional inactivation thus allowing detection of insert-bearing plasmids.

One notable non-shuttle vector is the so-called TRP1-R1 mini-plasmid[4], which comprises only yeast DNA and was made by circularising an *EcoR1* yeast fragment carrying the *TRP1* gene. It is fairly stable and is maintained in high copy number, but has not found much favour as a cloning vehicle. A second, specialist class of non-shuttle vectors comprises the linear ones designed for isolation of chromosome ends (telomeres)[5,6]. These will be described below.

The behaviour of a transforming plasmid in yeast, and thus its use in cloning, is wholly dependent on the sequences it carries. Broadly speaking there are two modes by which plasmids may effect transformation, the 'integrative' and the 'autonomous', and we shall now consider these in detail.

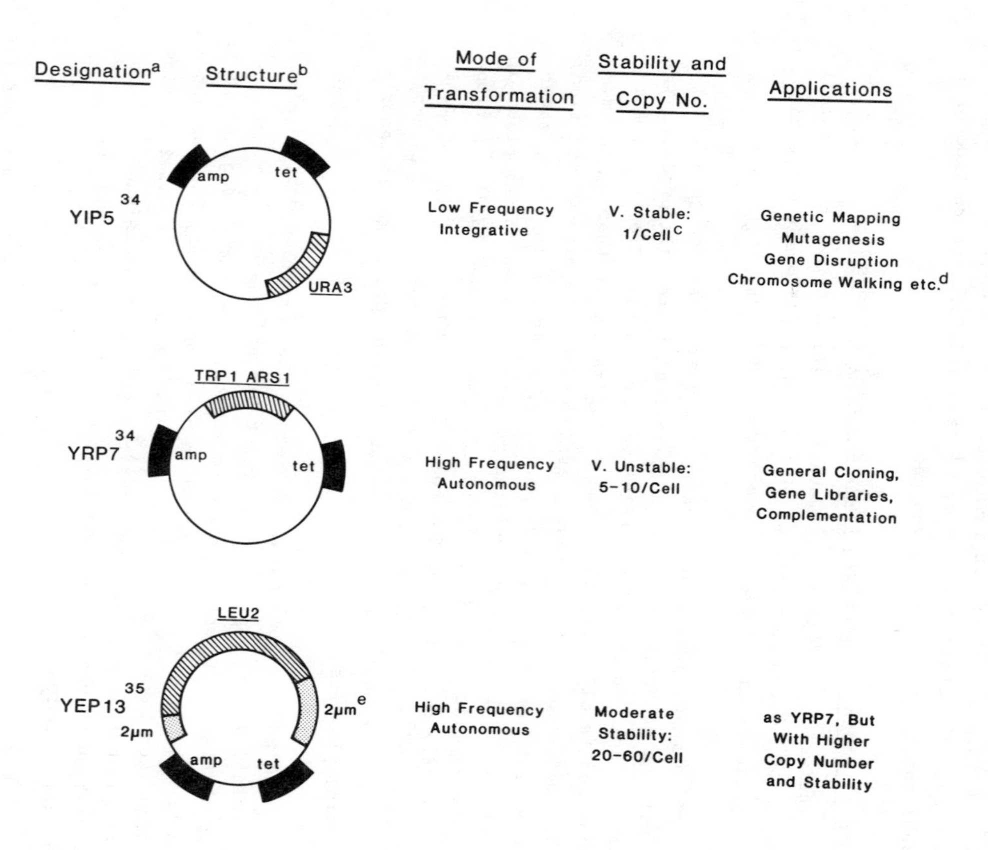

Designation[a]	Structure[b]	Mode of Transformation	Stability and Copy No.	Applications
YIP5[34]		Low Frequency Integrative	V. Stable: 1/Cell[c]	Genetic Mapping Mutagenesis Gene Disruption Chromosome Walking etc.[d]
YRP7[34]		High Frequency Autonomous	V. Unstable: 5–10/Cell	General Cloning, Gene Libraries, Complementation
YEP13[35]		High Frequency Autonomous	Moderate Stability: 20–60/Cell	as YRP7, But With Higher Copy Number and Stability

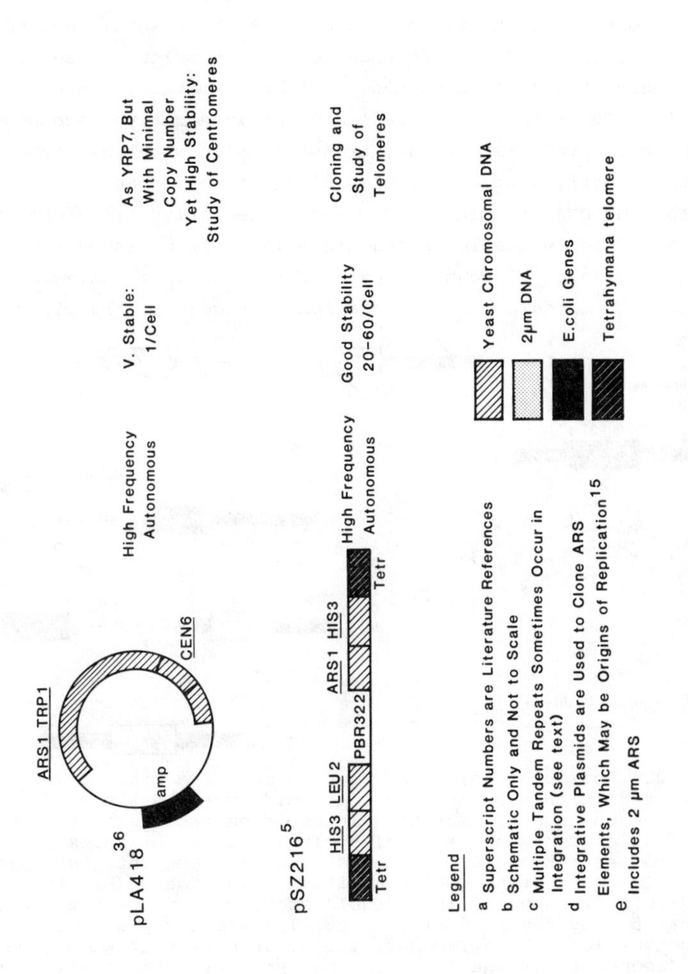

Fig. 1 Important Examples of Cloning Vectors in S. cerevisiae

Integrative vectors and their uses. Integrative vectors, designated "YIP" (Yeast Integrative Plasmids) can only achieve clonal expression (i.e. transformation) through integration into the genome by homologous recombination. They must therefore carry at least one region homologous to a yeast chromosomal sequence; they must also lack any "*ARS*" sequence (discussed below) whose presence would confer the autonomous mode of transformation. An example, YIP5, is illustrated in Figure 1. Since integrative recombination is a comparatively rare event, transformation with this type of vector is inefficient, and yields of <10 transformants per μg of transforming DNA are usual. However, integrative transformants are in general highly stable, a feature which distinguishes them from all other classes of yeast transformant. Possible consequences at the locus of integration are detailed in Figure 2. Multiple tandem

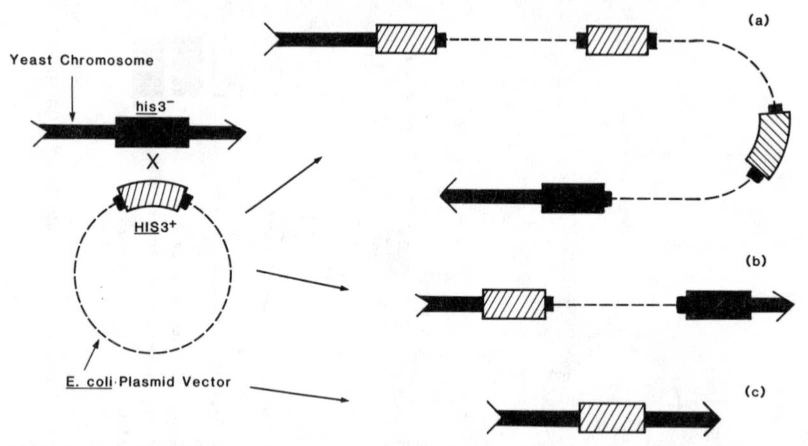

Figure 2. Possible consequences of integrative transformation, illustrated in this example by a plasmid carrying the wild-type *HIS3*$^{+}$ gene and a *his3*$^{-}$ mutant. Integration results from homologous recombination between the alleles. In up to 50% of cases multiple (sometimes as many as 10) copies of the gene are integrated (a). Another frequent event is the addition tranformant (b), the result of a single crossover. If a double crossover occurs, only one integrated copy remains (c), a situation which may also result from gene conversion. In all cases, the distribution of wild-type and mutant sequences in the integrant may vary from those illustrated, depending on the precise sites of the crossovers.

integrations of the yeast gene are common, but very often the primary product, the result of a single crossover, is of the addition type, resulting in a duplication of a stretch of yeast DNA flanking an integrated copy of the vector molecule. In fact, one of the flanks is the original chromosomal sequence - the other is a copy of the plasmid version. This addition product may be quite stable, or it may go through an excision event in which the inserted vector plus one of the yeast flanks is lost, leaving only a single copy of the yeast gene behind. A double crossover or a gene conversion event will have the same effect. In this case the integrated gene will have the full stability and function of a chromosomal gene.

In recent years, much attention has been paid to the use of "cut" vectors, which are plasmids either linearised by a single restriction enzyme cut at a site within a yeast sequence, or "gapped" by cutting with enzymes to remove a fragment from within the sequence[7,8].

The effects of these treatments on the transformation behaviour of the vectors are dramatic. Firstly, transformation frequency is increased many fold, and secondly, the enhanced integration occurs highly selectively at the chromosomal homologue of the cut sequence in the plasmid. This means that problems arising from reversion of host genes are diminished, and integration at sites other than the one of interest is, for practical purposes, eliminated. Thus a plasmid carrying several different yeast sequences can be effectively targetted to any one of them simply by cutting the plasmid in the gene of interest. The detailed consequences at the integration site are essentially the same with cut plasmids as with uncut ones, except that when a gapped plasmid is employed, the gap is repaired with chromosomal information, and both the sequences flanking the vector in the addition transformant are identical.

The use of these cut vectors has enormously enhanced the value of integrative transformation, the chief applications of which are as follows:

1. By targetting a vector carrying a selectable marker (e.g. *URA3*) to your favourite gene (YFG) that gene can be mapped by conventional means. Thus *any cloned sequence can be mapped*, even if no mutant phenotype is known[8,9].

2. If a shuttle vector carrying a wild-type yeast gene is targetted to integrate at a mutant allele on the chromosome,

appropriate restriction digestion and ligation of the total DNA of the addition transformant will generate a vector molecule carrying the mutant version of the gene. This can be directly transformed in *E.coli*, thus permitting *cloning of a mutant allele*[8,9]. This procedure works particularly well if a gapped plasmid is used, since in this case, provided the gap spans the site of the mutation, both the vector flanks will be copies of the mutant allele, and recovery of either version will give the desired result[8]. (As an additional twist, if the gapped plasmid is of the autonomously replicating variety (see below) gap-repaired replicating plasmids can be recovered directly, without the need for restriction and ligation[11]).

3. By appropriate choice of restriction enzymes, the DNA cut from the transformant can be made to carry chromosomal DNA adjacent to the targetted integration site. Ligation and cloning in *E.coli* in this case permits recovery of these adjacent sequences and, by repeated rounds of transformation, offers a relatively painless way of *chromosome walking*[9,12].

4. The fact that gap repair involves the precise chromosomal sequence spanning the gap means that when a mutation on the chromosomal allele is outside that region, a selectable wild-type allele will be generated. Provided enough gaps can be made, this gives a means of *fine structure mapping*[8].

5. If a mutant gene is available on a plasmid (and mutations may now be generated very specifically *in vitro*), it can be targetted to replace the wild-type allele. This gives a powerful route for *specific mutagenesis* of a strain without the problems inherent in exposing cells to mutagens[8,9,10,13].

6. A special version of (5) is the process of "gene disruption"[9,10,13,14], illustrated in Figure 3. A cloned gene has a selectable marker inserted *in vitro*, and this disrupted gene is targetted to replace one of the two wild-type alleles in a diploid. On sporulation, the disrupted allele will segregate, and if the gene is essential, 2:2 segregation for inviability will be observed. Thus this operation will *test if a sequence is essential*. Once again, no information about the function or phenotype of the gene in question is needed.

The above examples by no means exhaust the repertoire of tricks made easy by integrative transformation, but at least they

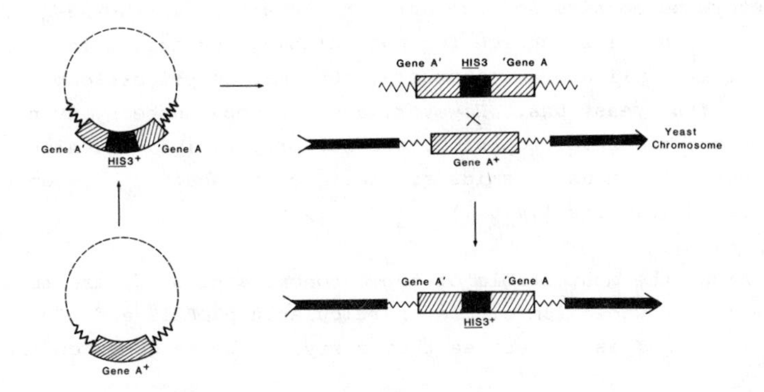

Figure 3. One step gene disruption[14]. A selectable gene such as *HIS3*+ is inserted by standard techniques into the gene to be disrupted, carried on an integrative plasmid. Transformation and selection of *HIS3*+ integrants yields cells bearing a disrupted copy of gene A+.

give some idea of its versatility. We shall now consider the second major class of vectors, those that replicate autonomously.

Autonomously replicating vectors. Vectors in this category differ from integrative ones in possessing a sequence known as an *ARS* (Autonomously Replicating Sequence). The *ARS*, which may be an origin of replication (for review, see [15]) allows otherwise integrative plasmids to be transmitted to progeny in an autonomously replicating form. Plasmids in this category induce transformation very efficiently (up to perhaps 20,000 transformants/µg DNA) but the transformants are all more unstable than their integrative counterparts. As an example, YRP7 (Figure 1) may be present in only a third of the cells of a transformant population, even under selection, and this proportion drops alarmingly rapidly on transfer to a non-selective medium.

ARS sequences may be found in most eukaryotic DNAs, but the important ones from the yeast cloners' point of view belong to two classes: (a) those derived from yeast chromosomal DNA, and (b) the *ARS* from the yeast 2µm circle, a naturally occurring high copy number plasmid. Vectors carrying chromosomal *ARSs* are designated "YRP" (Yeast Replicating Plasmids), while those built around the 2µm plasmid (or just its *ARS*) are usually termed "YEP" (Yeast Episomal Plasmid). The properties of both are similar and they may therefore be considered together. They are the "workhorses" of cloning - the first choice for most cloning operations in yeast. and since they replicate autonomously, they may in principle be recovered in bulk from yeast DNA. However, average copy numbers, even under selection, are never too high, and for preparative convenience therefore most autonomous plasmids are designed to shuttle between yeast and *E.coli* (Figures 1,4,6,7).

Among the most stable of the autonomous plasmids are those carrying the whole 2μm circle, particularly pJDB219 and pJDB248[1] (Figure 4). This is because they carry 2μm-borne genes coding for

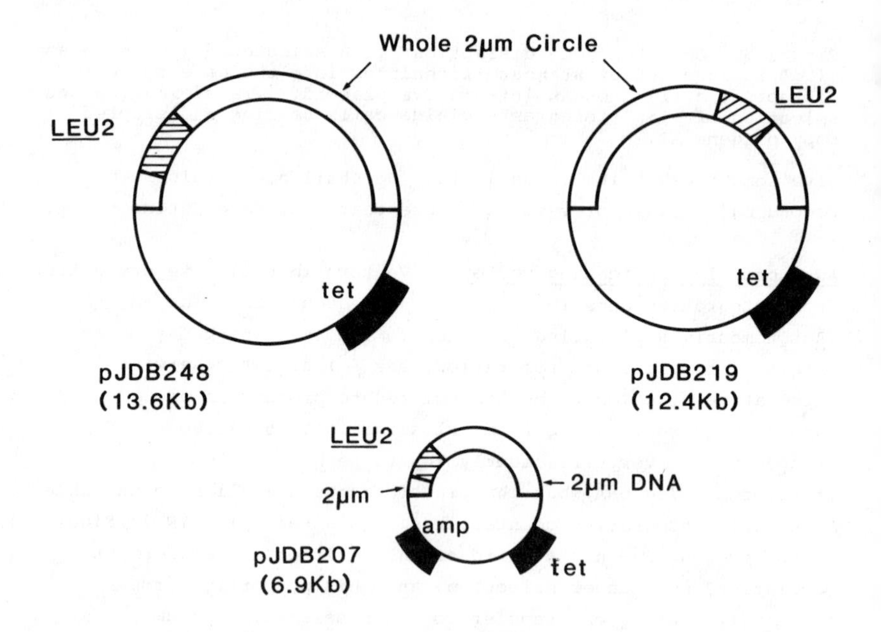

Figure 4. 2μm-based vectors. pJDB219 and pJDB248 are both "whole 2μm" vectors. pJDB207 is smaller and carries only part of the 2μm circle, to allow replication[1,21]. See text for further description.

products which help maintain stability,[16,17] and they are distributed uniformly amongst the cells in a population (copy numbers around 20 - 60). These plasmids may also be used to "cure" cells of endogenous 2μm circles by competition[18,19,20]. The resulting [cir°] cells provide a stable host for transformation with this type of "whole 2µm" vector, and are free of the risk of plasmid rearrangement resulting from recombination with endogenous 2µm circles. However, the "whole 2µm" vectors tend to be short of unique restriction sites and, being large, they are prone to re-arrangement and are not ideal for cloning large inserts. For these reasons, a smaller vector, pJDB207 (Figure 4), has been constructed. This uses only part of the 2µm circle, and is stable, transforms [*cir*$^+$] strains well, and has a fairly high copy number[21]. Like other 'partial 2µm' plasmids, it transforms [*cir*°] strains poorly or not at all, as the 2µm *ARS* requires the products of other 2µm-borne genes for efficient maintenance[15]. A comprehensive review on 2µm-based vectors is available[16].

Autonomously replicating plasmids can integrate by homologous recombination with chromosomal sequences, and the consequences in the case of the YEPs can be bizarre[22], since the 2µm plasmid carries inverted repeats and codes for an enzyme which promotes recombination between them[16]. However in practice, integration of either type of autonomous vector is sufficiently infrequent not to cause problems in cloning.

The extreme instability of the *ARS* vectors is due to their defective segregation, resulting in an average of 5 - 10 per cell, and very non-uniform distributions amongst the cells in a population. It can be overcome by the inclusion on the plasmid of a yeast centromeric sequence (*CEN*). First isolated by Clarke and Carbon[23], these sequences are presumed to give the plasmid a spindle attachment site which promotes efficient mitotic and meiotic segregation and thus, stability. In fact, following the first isolation of a *CEN*, others have been cloned in *ARS* plasmids by virtue of their stabilising effect. At the same time, by a mechanism not fully understood, the copy number of *CEN* plasmids is reduced to unity. Thus the *ARS*/*CEN* plasmid is sometimes referred to as a 'mini-chromosome'. However, it is by no means as stable as a normal linear chromosome. In fact it only does about as well as the "stabilised" version of the "whole 2µm" type of plasmid in a [*cir*°] background. On the other hand it segregates 2:2 at meiosis, whereas an *ARS* plasmid without a *CEN* is frequently lost at meiosis through instability. The stability of *CEN* plasmids is a technical convenience, and they tend to be more efficient in transformation than their *CEN*⁻ counterparts. They may also be advantageous in situations where a high intracellular content of a particular gene product resulting from the presence of the gene on a multi-copy plasmid might be deleterious. Various *CEN* sequences have been cloned and are available or have been built into vectors. One example is shown in Figure 1.

Autonomously replicating linear plasmids. Cut or broken linear DNA molecules tend to be unstable in cellular environments, and, as we have seen, are recombinogenic or liable to degradation. However, the ends of linear chromosomes are stabilised by specialised structures known as telomeres (Figure 5), which also permit correct replication of the ends of the molecule. It transpires that many lower eukaryotes have similar telomeric structures. Moreover in two protists - brewer's yeast and the trypanosomes which cause sleeping sickness, the telomere has been shown to undergo repeated

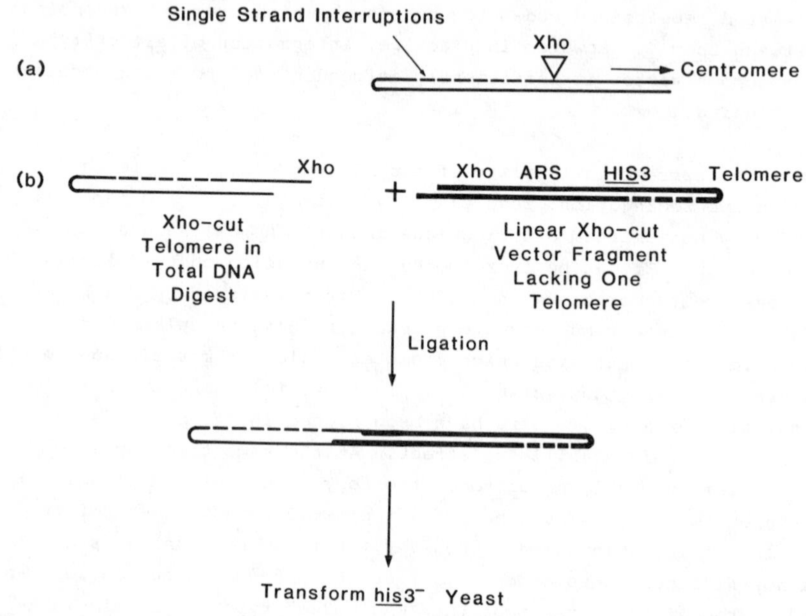

Figure 5. Schematic structure of yeast telomere and principle of cloning procedure. (a) simplified structure of telomere. (b) a linear vector is needed to clone liberated telomeres, since they only have one accessible end. The lines represent single DNA strands. In practice the cloning procedure may differ in detail from the principle illustrated[5,6].

cycles of growth and shortening, in a completely unexpected manner. For this and other reasons, telomeres are attracting considerable attention. In the case of yeast, autonomously replicating linear plasmids are available to permit the cloning of telomeres[5,6]. The archetype of this class (Figure 1) comprises telomeres from the ciliate protozoan *Tetrahymena* ligated to each end of a DNA molecule comprising a selectable yeast gene and at least one *ARS*. When the appropriately restricted plasmid is ligated with a restriction digest of yeast DNA and transformed into yeast, vectors that have acquired a functional yeast telomere(s) are capable of replication and transmission in an autonomously replicating mode (Figure 5). Interestingly, these plasmids are maintained in fairly high copy number and are at the top end of the stability range for autonomously replicating plasmids. Though of obviously rather specialised interest, they are finding increasing use in the study of telomere function in yeast and for the isolation of telomeres from other organisms.

Problems of Gene Expression and Some Solutions

From the foregoing, it will be apparent that expression of homologous yeast genes on autonomously replicating vectors presents no insurmountable difficulties, but this is not necessarily so for heterologous genes, i.e. genes from other organisms. One problem is simply that although some yeast genes possess introns, splicing signals and mechanisms are not apparently conserved throughout eukaryotes, so the processing of transcripts read from heterologous genes is not always accomplished successfully, and cDNA copies of genes may be needed to circumvent this problem.

There are of course two particular reasons for wanting to achieve expression of heterologous genes in yeast. One is that yeast carries certain eukaryotic-specific genes not found in prokaryotes. Provided the appropriate yeast mutants exist, complementation with random genomic DNA fragments from other organisms may provide a means of cloning the genes concerned. Secondly, yeast (especially brewer's yeast) provides an increasingly attractive system for large-scale production of heterologous gene products. It is extremely flexible. Expression can be maximised by using a high copy number plasmid carrying a strong promoter or, if this is undesirable, (for instance because the cells respond badly to a burden of foreign protein) insertion of a *CEN* sequence will maintain transformant stability while reducing plasmid copy number and the level of the gene product concerned.

Finally, being wholly non-pathogenic, brewer's yeast is thought to be relatively free of the risk of contributing undesirable toxic proteins to the product of interest, and with improvements in techniques for producing high yields and for secretion, so that contamination with intracellular proteins can be minimised, it is becoming increasingly popular.

Space precludes a detailed discussion of this rapidly developing area, but the reader may find a recent review on heterologous expression[24] of interest. A compendium of yeast vectors, including expression ones, is also available[25].

A recent improvement in this field has been the development of so-called "sandwich" vectors in which a 5' promoter is joined to a 3' downstream control sequence needed for proper termination of transcription[26]. In the example shown in Figure 6, the promoter carries a sequence coding for a few of the N-terminal amino acids

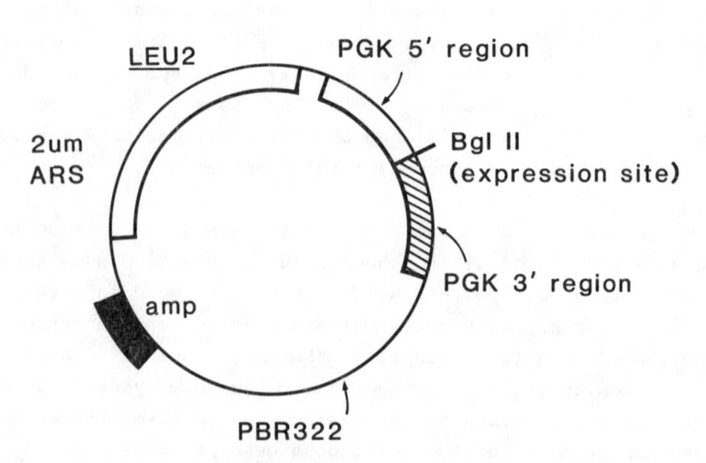

Figure 6. A sandwich expression vector PMA91[26]. PGK 5' and 3' sequences are derived from the yeast phosphoglycerate kinase. Sequences for expression are inserted at the unique *BglII* site.

of the protein with which it is normally associated. This allows efficient initiation of translation and is chosen so as not to interfere with the function of the desired product. The plasmid illustrated in Figure 6 has been used successfully for the expression in yeast of a number of proteins including interferon and chymosin[26,27].

Secretion. Secretion of polypeptides through cell membranes requires the presence on their amino-terminal ends of special peptides which signal the cell that the protein is for export, and which are cleaved from it as part of the secretory process. To achieve secretion of your favourite gene product therefore, a nucleotide sequence specifying such a leader must be engineered onto the 5' end of your plasmids' coding sequence in such a manner that the junction between the two sequences is correctly processed. More than one candidate for this purpose is emerging, but a popular choice at the moment is the yeast alpha-hormone. This is a small polypeptide (17 amino-acids) involved in yeast mating reactions, and normally secreted into the cells' environment. The genes specifying this hormone have been cloned and have been successfully used to achieve secretion of yeast invertase[28], human epidermal growth factor[29], interferon and β-endorphin[30]. How versatile alpha-factor will prove in the long run for this purpose remains to be seen.

LacZ gene fusions. The use of so-called fusion proteins and plasmids is an aspect of recombinant DNA work which has long been in use with *E.coli*, but has recently become widely available, in all its aspects, to those working with *S.cerevisiae*. This is a large subject which can only be dealt with superficially here, and for detailed information on yeast applications, the reader is referred elsewhere[31,32].

Far and away the most important fusion system in yeast, as in *E.coli*, is the one built around the Z gene of the *E.coli lac* operon. This gene codes for β-galactosidase in *E.coli*, but functions also in yeast, and in both organisms this enzyme activity is readily assayed in either solid or liquid media. An additional key property of β-galactosidase is that the first 30 N-terminal amino-acids can be deleted without loss of function. Moreover, mutant forms of the gene (designated '*lacZ*) have been isolated which only generate functional enzyme when the deleted sequence has been replaced with a stretch of amino-acids which may be quite long and unrelated to the β-galactosidase gene itself. This provides one obvious means of cloning coding sequences and promoters. For instance, the promoter of your favourite gene, plus a few of its N-terminal amino-acid codons, may be fused, in frame, to the 5' end of the truncated '*lacZ*- sequence on a plasmid equipped for replication in yeast. Provided the promoter is functional, its operation will now generate a fusion protein which, despite its hybrid nature, has assayable β-galactosidase activity. Since this activity is now under the control of the inserted promoter, the genetic and molecular regulation of that promoter is immediately open to exploration, and the study of gene regulation is one of the main applications of fusion plasmids. If the plasmid construct already includes an appropriately sited promoter, the fusion plasmid allows the cloning of any sequence carrying an open reading frame.

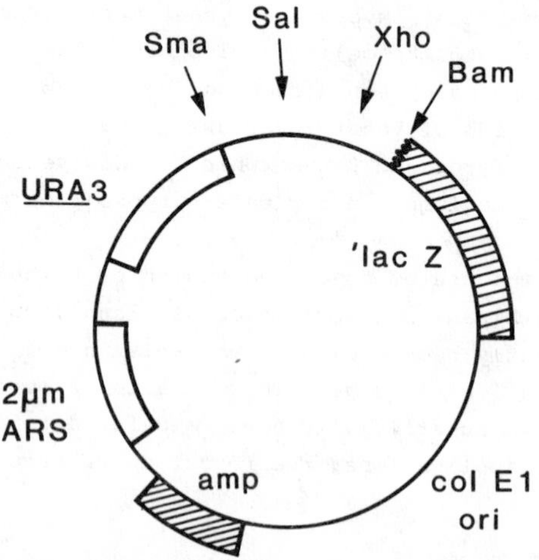

Figure 7. Shuttle vector pLG670-Z used for constructing *lacZ* gene fusions for study in yeast[31]. Any of the unique restriction sites shown may be used to engineer the in-frame insertion of a sequence of reading frame DNA. Provided the insert carries a promoter, functional fused β-galactosidase will be generated in yeast.

It also provides a useful avenue for the purification of polypeptides for use as antigens. In this case, either the enzymic activity of the β-galactosidase moiety of the fusion protein, or antisera against it, may be used as a purification tool.

In yeast the fusion systems have considerable flexibility. Levels of protein may be increased by the use of multi-copy 2μm-based plasmids and strong promoters, or reduced by *CEN* plasmids and less active promoters. Repressible or inducible promoters may also be used to control the timing of expression of the fused gene. One example of a *lacZ* fusion plasmid is illustrated in Figure 7.

The main problem encountered in using fusion plasmids is the need to get the inserted coding sequence in-frame with the truncated *lacZ* gene. If the sequence of the insert is known, this may be relatively easy, but in the absence of such data, one may have to revert to cloning random *Bal31* deletions, often using blunt end ligation and synthetic linkers. Discussion of this type of strategy may be found elsewhere[31,32].

Conclusion

This has been a very sketchy introductory survey of what has become a very large and fast-growing area of molecular biology, and reference has already been made to articles dealing with particular aspects in more detail. The interested reader may find another general review helpful[33]. It is this author's hope that this brief outline has to some extent illustrated the versatility of yeast as a cloning vehicle, and that it may serve as a springboard for those wishing to dive deeper into the pool.

References

[1]A.Hinnen, J.B. Hicks and G.R. Fink, Proc. Natl. Acad. Sci., 1978, 75, 1929.
[2]J.D.Beggs, Nature, 1978, 275, 104.
[3]H.Ito, Y.Fukuda, K. Murata and A.Kimura, J.Bact., 1983, 153, 163.
[4]V.A.Zakian and J.F. Scott, Mol. Cell. Biol., 1982, 2, 221.
[5]J.W.Szostak and E.H. Blackburn, Cell, 1982, 29, 245.
[6]J.W.Szostak, Methods in Enzymology, 1983, 101, 245.
[7]T.L. Orr-Weaver, J.W. Szostak and R.J. Rothstein, Proc. Natl. Acad. Sci., 1981, 78, 6354.
[8]T.L. Orr-Weaver, J.W.Szostak and R.J. Rothstein, Methods in Enzymology, 1983, 101, 228.
[9]F.Winston, F.Chumley and G.R. Fink, Methods in Enzoymology, 1983, 101, 211.
[10]S.Scherer and R.W. Davis, Proc. Natl. Acad. Sci., 1979, 76, 4951.
[11]T.L. Orr-Weaver and J.W.Szostak, Proc. Natl. Acad. Sci., 1983, 80, 4417.

[12] J.I. Stiles, Methods in Enzymology, 1983, 101, 290.
[13] D.Shortle, J.E. Haber and D.Botstein, Science, 1982, 217, 371.
[14] R.J. Rothstein, Methods in Enzymology, 1983, 101, 202.
[15] D.H. Williamson, Yeast, 1985, 1, *in press.*
[16] J.R.Broach, Methods in Enzymology, 1983, 101, 307.
[17] Y.Kikuchi, Cell, 1983, 35, 487.
[18] M.Dobson, A.B. Futcher and B.S. Cox, Curr. Genet., 1980, 2, 201.
[19] E.Erhart and C.P. Hollenberg, Curr. Genet., 1982, 3, 83.
[20] A.M. Cashmore, PhD Thesis, N.I.M.R., Mill Hill, 1984.
[21] J.D.Beggs,"Molecular Genetics in Yeast", Munksgaard, Copenhagen, 1981, p.383.
[22] S.C.Falco, Y. Li, J.O.Broach and D.Botstein, Cell, 1982, 29, 573.
[23] L.Clarke and J.Carbon, Nature, 1980, 287, 504.
[24] S.M. Kingsman, A.J. Kingsman, M.J. Dobson, J. Mellor and N.A. Roberts, Biotech. and Genet. Engineering Rev., 1985, 3, *in press.*
[25] S.A. Parent, C.M. Fenimore and K.A. Bostian, Yeast, 1, *in press.*
[26] J.Mellor, M.J. Dobson, N.A. Roberts, M.F. Tuite, J.S. Emtage, S.White, P.A. Lowe, T. Patel, A.J. Kingsman and S.M. Kingsman, Gene, 1983, 24, 1.
[27] S.M. Kingsman and A.J. Kingsman, Interferons : From Molecular Biology to Clinical Application, 1983, Soc. Gen. Microbiol. Symp., 35, 211.
[28] S.D.Emr, R.Schekman, M.F. Flessel and J. Thorner, Proc. Natl. Acad. Sci., 1983, 80, 7080.
[29] A.J. Brake, J.P. Merryweather, D.G. Coit, V.A. Heberlein, F.R. Masiary, G.T. Mullenbach, M.S. Urdea, P. Valenzuela and P.J. Barr, Proc. Natl. Acad. Sci., 1984, 81, 4642.
[30] G.A. Bitter, K.K. Chen, A.R. Banks and P.Lai, Proc. Natl. Acad. Sci., 1984, 81, 5330.
[31] L.Guarente, Methods in Enzymology, 1983, 101, 181.
[32] M.Rose and D.Botstein, Methods in Enzymology, 1983, 101, 167.
[33] D.Botstein and R.W.Davis, "The Molecular Biology of the Yeast *Saccharomyces* : Metabolism and Gene Expression", Cold Spring Harbor, Cold Spring Harbor, New York, 1982, p607.
[34] K.Struhl, D.T. Stinchcomb, S.Sherer and R.W. Davis, Proc. Natl. Acad. Sci., 1979, 76, 1035.
[35] J.R.Broach, J.N. Strathern and J.B. Hicks, Gene, 1979, 8, 121.
[36] L.Panzeri and P.Philippsen, EMBO Journal, 1982, 1, 1605.

5
Cloning in Mammalian Cells

By R. E. Spier

DEPARTMENT OF MICROBIOLOGY, UNIVERSITY OF SURREY, GUILDFORD, SURREY GU2 5XH, U.K.

1. Introduction

The relationship between molecular biology and biotechnology may be most easily discerned from a juxtapositioning of the definitions of these two subject areas. Taking biotechnology as the activity which leads to the conversion of raw materials to final products, wherein either the raw materials and/or an entity involved in the transforming process has a biological origin, and defining molecular biology as that body of knowledge and understanding associated with the interactions of molecules in living systems it would seem that there would be little that these two separate areas have in common. Yet it has been through our increased understanding of molecular biology that we have developed capabilities for the directed manipulation of the genomes of all classes of living organisms. The result of that ability has been the deliberate engineering of particular biological entities which can be used either in the manufacture of new bioproducts or in the improvement of the efficiency of generation of more traditional materials. This new found faculty has engendered a reexamination of the way in which bioproducts should be made in the future. In short it has become possible to produce almost any particular protein, (as a linear sequence of amino acids (cf. below)), in prokaryotes or such diverse eukaryotes as yeast cells, animal cells or plant cells. The task which remains to be done is to determine for each and every such product the most cost-effective system for manufacture, having included in such consideration the relative benefit to risk ratios and the requirement and modes of operation of those social bodies which provide the "license to produce" certificates.

This chapter is about those particular methods which can be used to generate proteinacious materials by using animal cells in culture as the primary production agency. Insufficient work has been done to be confident that the dominant production technology for any one particular product will, having evaluated the alternatives, be based on animal cells in culture. Yet from considerations discussed below it is also unlikely that all manufacturing processes will use cells other than animal cells as their biological workhorses. The next decade or two should therefore see a shaking out of the alternative production technologies which provide maximised efficiencies for particular products. It would not be unreasonable to expect that animal cell systems and their genetically engineered counterparts will have a role to play in that technological future.

2. Factors which affect system productivity

The productivity of a biotechnological system depends on the interaction of a number of key factors. Ten such factors have been listed in Table I yet others could have been added. Suffice to relate that this represents a complex situation which is rarely dealt with as a whole. Most attempts at system optimisation involve a series of successive approximations or iterations round the loop of a series of these factors. However, the most relevant conclusion from this situation is that, having obtained a cell which has been genetically engineered to dedicate its 'all' to the expression of a particular protein the subsequent treatment of such a cell in the delineation of the other process and technology parameters can either completely suppress productivity or enable it at maximal levels. By contrast the most efficient technological environment will not necessarily lead to the most extensive productivity if the cell is inappropriate. For the most advantageous system therefore the cell, the process and the technology have to be optimal.

3. Animal Cells as Biological Substrates for Manufacturing Processes

Animal cells in culture are often regarded as being "delicate", subject to disruption by shear forces, slow growing and poor sources of product. Indeed, when compared to systems based on bacteria or yeasts, animal cells would appear to be at a severe disadvantage. (Table II). This situation can be illustrated by comparing the biomass productivity of a bacterium growing in a continuous culture with a doubling time of 1/3 hr compared with an animal cell continuous culture with a doubling time of 12 hours. The bacterial

Table I

Factors which affect the productivity of biotechnological systems

Process Factors

- The cell
- The medium
- The product
- Assay systems
- Time/intensity relationships

Technology Factors

- System reliability
- Ability to scale-up efficiently
- Ease of handling (Separations)
- Biomass concentration
- Diffusion dependent activities

Table II

Features promoting the use of Bacteria or Yeast vis à vis animal cells in culture

- Higher biomass productivities
  (wt biomass/vol of reactor/time)

- More reliable production systems
  - less fastidious medium requirements
  - easier control of contamination

- Cheaper media

- Can be engineered to produce proteins which have been traditionally produced in animal cells.

- Storage and resuscitation of cells simple

culture will produce about 50 times more biomass per unit time compared with the animal cell culture. It is interesting to note that the surface area to volume ratio of the bacteria in the bacterial culture is only 10 times that of the animal cell. There is thus an opportunity for a more efficient animal cell culture system (Table III). This issue will be dealt with in more detail in a later section (v.i. Section 5).

Table III

Biomass Productivities of Bacteria and Animal Cells in Culture Compared

Parameter	Cell Type	
	Bacterial Cell	Animal Cell
Cell Diameter (μm)	1	10
Surface area (cm^2)	3.1×10^{-8}	3.1×10^{-6}
Volume (cm^3)	5.2×10^{-13}	5.2×10^{-10}
SA/volume	59×10^3	5.9×10^3
Doubling time (hr)	0.3	12
Cell concentration ml^{-1}	5×10^9	4×10^6
Productivity No. of cells/L/hr	7.5×10^{12}	1.7×10^8
Productivity Gm cells/L/hr	3.75	0.085
Relative Productivity	44	1
Relative SA/vol	10	1

However the productivity of a system is not generally determined by biomass alone (except perhaps for single cell protein producing systems). The biological activity of the product and its ability to perform in an environment which reacts strongly and negatively to the presence of foreign bodies is a most relevant consideration. These latter factors are particularly germane when one reviews the advantages and disadvantages of producing materials from genetically engineered bacteria or yeasts or of producing the same materials from animal cells. In Table IV such differences are lsted. Clearly, all of the factors listed in Table IV do not prevail in any one situation and indeed useful materials can be made in bacteria (insulin, α-interferon) and yeast (Hepatitis B surface antigen) yet it is important to draw attention to regulatory agency attitudes as such considerations influence to a considerable degree the area of "the practicable". Thus the presence of transforming DNA (in a latent, active or putative form) requires much testing to ensure safety. Also the presence of endotoxins, pyrogenic factors or allergins will prevent products reaching the

Table IV

Problems arising in the production of materials from genetically engineered bacteria

- regulatory agency attitudes
- products retained within the cell
- products toxic to producing cell
- difficulty of and need for endotoxin removal
- damage to products from intracellular and extracellular protease
- bacteriophage contamination
- plasmid stability
- biological activity of product (enzyme and antigen)
- imperfect spatial configuation leading to recognition as foreign
- absence of appropriate post translational modification systems
  - glycosylation
  - carboxylation
  - hydroxylation
  - phosphorylation
  - disulphide bridge formation
  - signal peptides inappropriate for:-
    - excretion
    - membrane insertion
    - for nuclear processing
  - end region definition - fusion protein problems

market place. Finally, it is clear that it is in the area of post translational processing that major and activity determining events occur. Not only does the animal cell provide the appropriate and specific enzymes and cofactors but it also provides a structured cytosol in which the proteins fold up and form the native disulphide bonds which enable the molecule to express the unique three dimensional structure which characterises the enzymic, hormonal and immunological reactivity of the molecule. Such factors have led to a redoubled effort to enhance the productivity of materials which can be made from animal cells. The method of choice to achieve this end is to genetically engineer the animal cell to become an "over producer" of a particularly desired commercially valuable material.

4. Genetically Engineered Animal Cells

Animal cells in culture have formed the basis for the production of virus vaccines for some 35 years since Enders discovered in 1949 that Poliovirus could grow in primary green monkey kidney cells in roller tube cultures. The step forward to genetically engineered animal cells began with Harris's work in the 1960's showing that the fusion product of a human and mouse cell could be viable which was followed in 1975 with Kohler and Millstein's breakthrough in producing a fused lymphocyte-myeloma mouse cell line which retained its viability and yet excreted antibody molecules. The era of "hybridoma" technology for the production of monoclonal antibodies (McAb) had arrived. Since that date some 200 Companies, worldwide, have or are using similar procedures to derive McAb producing cells as such antibodies have commercial value in a variety of systems (Table V).

In addition to the method of animal cell fusion to produce "genetically engineered" cell lines there are a wide variety of alternative techniques some of which have resulted in the production of new cell lines for proteins which are likely to be manufactured from such cells in the commercial areas. Table VI lists such putative commercial products based on the exploitation of genetically engineered animal cells in culture. It should be noted that such materials are glycosylated (β & γ -interferons) and Blood Clotting Factor IX has 12 glutamate specific carboxylations (vitamin K dependent) and one hydroxylation on the β-carbon of an aspartic acid residue. As such cells have a clear commercial potential it would be appropriate to review the methods by which they can be produced.

Table V

Uses of Monoclonal Antibodies

- Diagnosis of disease
- Medical imaging
- Therapy (site specific toxins)
- In science
  - cellular antigens
  - cytoskeleton
  - antigen/immunogenic structure analysis
- In epidemiology
  - shift in epitope type
  - movement of disease
- Industrial processes based on immuno adsorptive separations (interferon)
- Immunoactive assay systems

Table VI

Proteins likely to be made from genetically engineered animal cells

- Tissue plasminogen activator
- Blood clotting factors
  - VIII
  - IX
- Hepatitis B surface antigen
- Interferons
  - β
  - γ
- Human growth hormone

4.1 Vector Systems

There are many vector systems which can be used to transform animal cells. A list of such vectors is presented in Table VII. Both the SV40 and polyoma vectors have been used as lytic vectors in cells which harbour in a constitutive fashion viral genes which complement for the genes which have been removed from the transducing vector. However, naked DNA, presented as a calcium phosphate precipitate, is still a popular method of transfection particularly when this is coupled with genes which can act as selective markers and carrier DNA (often from salmon sperm) which increases the efficiency of the process.

Table VII

Vector systems for animal cells

- DNA + selective marker + carrier DNA
- SV40 + cell with constitutive and complementary genes
- Polyoma virus
- Bovine Papilloma virus
- Herpes viruses
- Retroviruses

4.2 Selective Markers

The selection of the cell which has taken up the DNA is a technology which is developing rapidly. A variety of markers are available which can be incorporated into an engineered vector or which can be added to the gene to be transferred prior to forming the calcium phosphate precipitate (Table VIII).

Table VIII

Selective Markers

- Thymidine kinase
  (for use with ThK$^-$ cells)

- Dihydrofolate reductase
  (for use with DHFR$^-$ cells or
  cells inhibited with methotrexate)

- Aminoglycoside -3'-phosphate transferase
  (for use with cells inhibited
  by the antibiotic G418)

4.3 Gene Transfer Methods

To improve the efficiency of the uptake of the DNA a variety of methods have been developed. While the calcium phosphate method is a rather random process, the use of microinjection is a direct method which deposits DNA directly into the nucleus. The method is technically demanding yet once learned is a valable adjunct to an animal cell genetic engineering program. Table IX depicts these and other methods which are currently in use for transfection work.

4.4 Recipient Cells

As in any technology, certain manifestations dominate the field. A number of cell lines have been used the most common of which are listed in Table X. In addition to those in the table Baby Hamster Kidney cells have been shown to express Blood Clotting Factor IX after having been microinjected with the appropriate gene.

4.5 Positive regulatory sequences and enhancers

Positive regulation occurs for many genes. The sequences which respond to a rapid heat shock, change in metal ion concentration or to the presence of glucocorticoid hormones are all found upstream of the sites involved in the initiation of mRNA transcription. Such control sequences can be used in

Table IX

Gene Transfer Methods

- DNA + DEAE Dextran
- DNA + Carrier DNA + Calcium phosphate
- Lipid Vesicles
- Bacterial Protoplasts
- Microinjection

Table X

Recipient Cells

- Mouse LM (TK^-)
- Chinese Hamster Ovary Cells ($DHFR^-$)
- Mouse 3T3 cell line
- Fertilised oocytes of
    - mouse
    - rabbit
    - sheep
    - pig

constructs to develop effective processes based on two phase reaction systems. In the first phase the cells will be grown, while in the second phase growth will be repressed (by the removal of growth factors) and the gene for the particular bioproduct will be activated through the modulation of the activity of the regulatable positive elements which have been incorporated in the constructs.

Enhancers, which are DNA sequences that can "enhance" the expression of nearby genes do not however have to be located upstream of the gene they are affecting and they are not orientation specific. Enhancers can have a positive effect even if placed several kilobases from the gene whose expression they are stimulating. Although not gene specific, some enhancers are more effective in certain heterologous cells than others, and the same can be said for other positive control elements and promoter regions. For maximum expression the control sequences have, to some extent, to be matched to the host cell in which the constructs are to be propagated. Enhancers normally increase gene expression in the absence of a particular stimulus and are of value in the development of one stage systems with high constitutive levels of gene expression (Figure 1).

FIGURE 1

REGULATION OF EXPRESSION

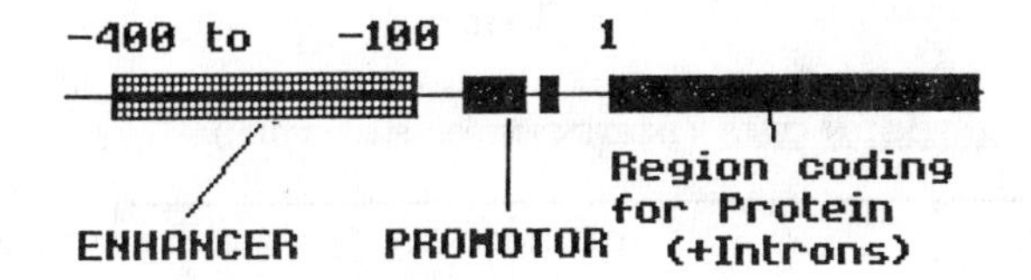

1. copy number
2. transcription
3. translation

5. New Developments in the Technology of Animal Cell Systems

To match the developments in genetic engineering, technological developments in the way in which animal cells may be exploited more reliably and efficiently have been developed. Figure 2 demonstrates that cells which require surfaces to grow on can be produced in systems which scale-up by increasing the number of multiples (multiple process systems) or by increasing the size of the unit (unit process systems). In the latter case a number of alternatives have been investigated as shown in the Figure. Suspension cell systems also have been developed recently in that more reliable sensors for pH and dO_2 are now routinely used in conjunction with computer based monitoring and control systems. Liquid and gas flow rates can be calculated more accurately and the total usage, and the rate of usage of system modulating materials (acid, base, air) can be readily obtained from the computer printout.

FIGURE 2

ALTERNATIVE SYSTEMS FOR THE LARGE-SCALE PRODUCTION OF ANIMAL CELLS

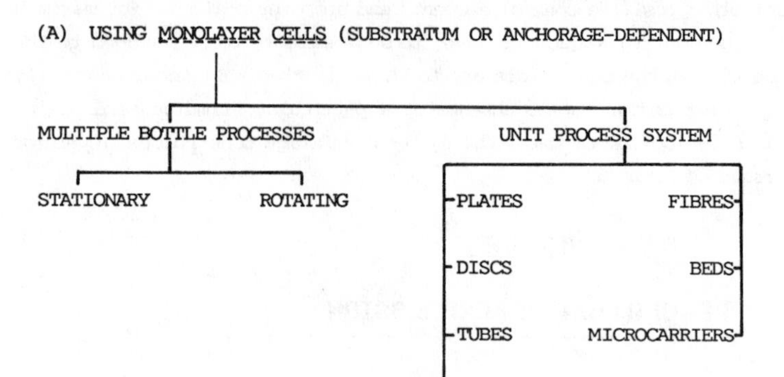

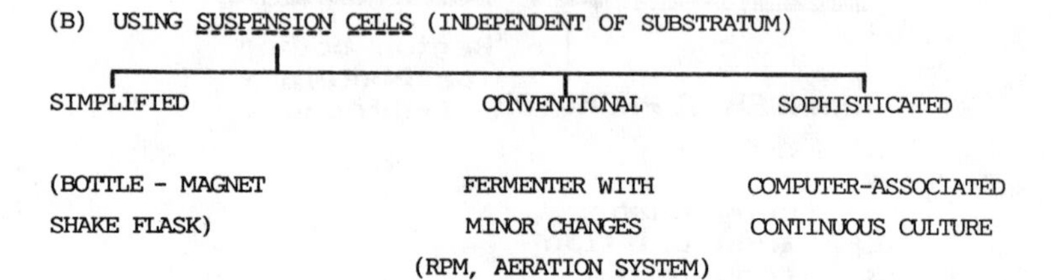

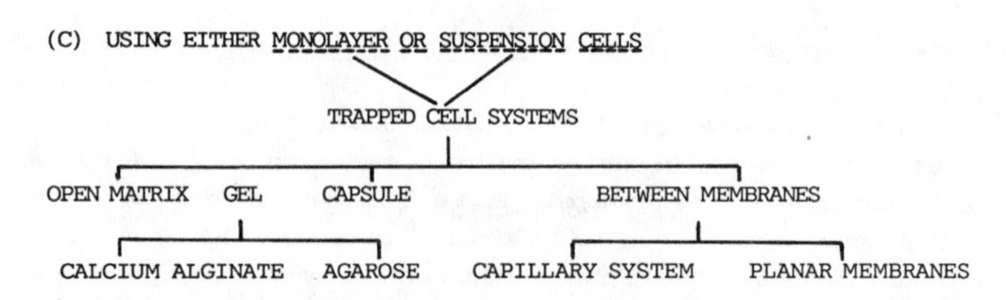

More recently yet, systems based on trapped cells have shown considerable promise. Three alternative strategies may be discerned. Gell entrapment of cells in either beads or slabs have been assayed and commercial systems are on offer to exploit cells trapped in a more or less open matrix, in spherical capsules made from polylysine and in systems which hold cells in compartments made from semipermeable or filter membranes where the latter can be in the form of capillary tubes or planar membranes.

In such trapped cell systems it is possible to grow and exploit animal cells at local cell concentrations of between 10^8 to 10^9 cells per ml. Cells at these concentrations have a consistency closely resembling that of a paste and indeed, were the cells undeformable 10 μm diameter spheres, then the packed cell material would have about 2 x 10^9 cells/ml. However as animal cells are deformable (unlike yeast and bacteria which are bounded by a rigid cell wall) it is theoretically possible to obtain animal cells so deformed as to occupy all the space available, in which case a cell concentration of 4 x 10^9 cells/ml is posible. Clearly, the most important consideration becomes the ability to supply nutrients and remove toxic and non-toxic products of metabolism from such cell pastes or semi solid masses. In practice this is achieved by circulating the growth or maintenance medium in the inner lumen of the capillary or the alternate space between membranes for the planar system. (Fig 3, Fig 4). Using the human body as an analogy it can be calculated that particular cells may be serviced by circulating fluids providing that they are within about 0.4mm or 40 cells diameters of such fluids. However <u>in vitro</u> studies to support such a contention have yet to be described.

FIGURE 3

CAPILLARY REACTOR

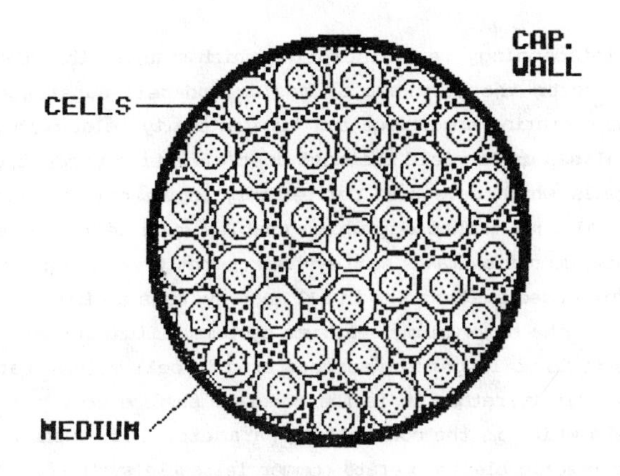

FIGURE4
MEMBRANE REACTOR.

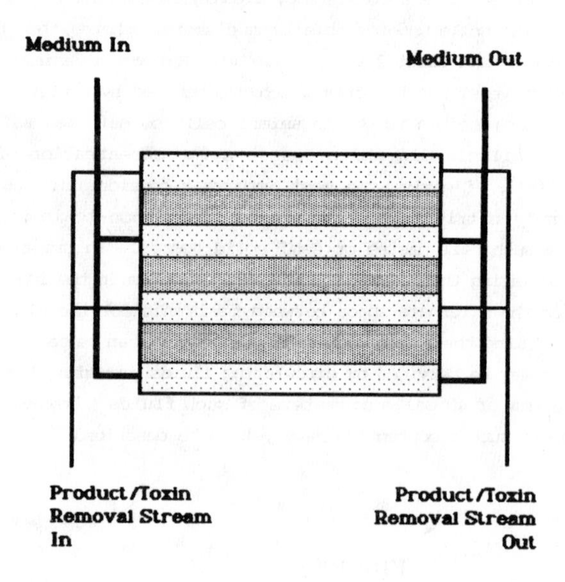

6. Conclusion

Process Biotechnology is a discipline which requires that the physical hardware of a system the firmware of medium and cell substrate and the software of the operating conditions and time/intensity relationships are so defined and realised as to yield the most cost/effective production system. Using animal cells which have been deliberately engineered to overproduce a required material in a technology which enables such cells to be used at maximised biomass densities makes such systems competitive with prokaryotic or lower eukaryotic based systems for in all such systems a similar amount of biomass is held in the same volume. In such systems diffusion characteristics are similar as it is no longer the surface area to cell volume ratio which controls the reactivity; rather it is the membrane surface area per volume of packed cell mass which is the controlling parameter. If, then, it becomes technologically practicable to operate commercial scale systems in this mode then factors other than the technological determine the value of the system.In producing materials for animal prophylaxis, therapy or physiological enhancement the animal cell based system has an additional edge in that it offers the 'natural' environment for the production of such materials which therefore express the native biological activity. Also such materials can be produced in a concentrated form in a simple and inexpensive medium as a result of recent activities which have led to a number of low protein serum substitutes. For these reasons animal cells may be adopted as a vehicle of choice for the cloning of useful genes as the putative and practical technological limitations of animal cells in culture are well on the way to being eliminated.

Suggested Further Reading

1. Cloning in Mammalian Cells
Ch 9 in Principles of Gene Manipulation
R.W. Old and S.B. Primrose. Blackwell
2nd Ed. 1981 pp 121-137.

2. Eukaryotic Viral Vector
Ed. Y. Gluzman
Cold Spring Harbour Laboratory 1982, pp 1-216.

3. Enhancers and Eukaryotic Gene Expression
Ed. Y. Gluzman and T. Shenk
Cold Spring Harbour Laboratory 1983 pp1-218

4. Gene Amplification
G.R. Stark and G.M. Wahl
Ann. Rev. Biochem. 1984 447-491

5. Genetic Engineering of Animal Cells
F. Colbere-Garapin & A.C. Garapin
Ch 16 in Vol II (pp 405-429) in
Animal Cell Biotechnolgy
Ed. R.E. Spier & J.B. Griffiths
Academic Press 1985
(Also for further aspects of cell technology)

6. Expression of Amplified Hepatitis B virus
Surface antigen genes in Chinese Hamster Ovary cells.
M.-L. Michel, E. Sobczak, Y. Malpieve P. Tiollais & R.E. Streecki
Biotechnology Vol 3, June 1985 561-566

7. Expression of active human clotting factor IX from
recombinant DNA Clones in Mammalian cells
D.S. Anson, D.E.G Austen & G.C. Brownlee
<u>Nature</u> <u>315</u> 1985 683-685.

6
The Genetic Manipulation of Crop Plants

By B. J. Miflin

BIOCHEMISTRY DEPARTMENT, ROTHAMSTED EXPERIMENTAL STATION, HARPENDEN, HERTS. AL5 2JQ, U.K.

I. INTRODUCTION

A tremendous upsurge of interest in plants has arisen as a result of the potential that now exists for genetic manipulation. This has led to a large effort in this subject within the AFRC, to the founding of many new commercial companies in the USA, and now even in Britain, and to the buying up of seeds firms by large multi-national companies. It is therefore relevant to ask what is this new agricultural biotechnology about, what are its aims, its methods and its success to date, and what are the potentials for the future? I hope to deal with these questions as well as possible within the present limitations.

II. DEFINITION

My definition[1] which agrees loosely with a commercial definition of agricultural biotechnolgy is as follows:

(1) Novel techniques for altering the genetic composition of plants and microorganisms associated with them in agriculture, including recombinant DNA techniques, limited gene transfer, partial genome transfer, somatic hybridization, somacloning and allied tissue culture techniques but excluding conventional plant breeding.

(2) Application of techniques for _in vitro_ production, multiplication

and selection of plants including tissue culture, protoplast culture, plant regeneration, and the selection of mutant cells or plants by novel techniques.

III. AIMS

The aim of genetic manipulation with respect to crop plants can be stated in general terms as..'the creation of new genotypes of crop plants which are one or all of the following: more productive, more efficient in resource use, and more resistant to pests, diseases and adverse environments.' It is implicit in this definition that such plants will be produced by means other than sexual crossing. However, it should be stressed that genetic manipulation has gone on in conventional plant breeding programmes for a long time and that it has been oustandingly successful. In the future development of agriculture it is likely that novel and existing techniques will progress hand-in-hand and not as separate competing approaches. The grandiose general aim stated above is likely to be advanced by setting up and achieving a series of subsidiary aims. Currently it is possible to identify a number of these:-

(a) Analysis of Plant Function

Recombinant DNA technology is the most powerful analytical technique ever made available to plant scientists - allied to a greater use of plant genetics it will provide a means for understanding the mechanisms and control of many plant processes. To achieve this scientists are concentrating on:-
(i) The isolation and structural analysis of individual genes. So far less than 100 of the 20,000 or so genes present in plants have been isolated and sequenced. Nevertheless a large amount of information of plant gene structure has been obtained. In general plant genes are constructed in similar ways to those of other eukaryotes. However, one surprise has been the complexity of the genetic control for certain apparently simple enzymic

steps (e.g. the multiplicity of genes for the small subunit of RuBP carboxylase[2] or for glutamine synthetase[3]).

(ii) Identification of mechanisms controlling gene expression. Plant scientists are using similar techniques to those pioneered in animals to identify and analyse the sequence importance in obtaining regulated gene expression and some success is being achieved (see below).

(iii) Development of transposon tagging. Although transposable elements were first postulated by McClintock on the basis of experiments with plants they have been studied in minute detail in bacteria. In the bacterial system they have proved useful in many ways but not least in gene identification. Essentially, when they insert themselves into or adjacent to a gene they inactivate it. This allows mutant phenotypes induced by transposon insertion to be traced back to a particular gene and thus the eventual isolation of the wild type gene governing the phenotype of interest. Transposable elements have now been isolated and sequenced from several plants. Subsequently they have been used to isolate genes for certain phenotypes observed in maize[4,5,6]. It should eventually be possible to extend this approach to other crops and a wide range of traits in order to isolate the genes governing them.

(b) Defined Gene Transfer

The eventual aim is to be able to insert any gene into the germ line of any plant species and to have the gene expressed in a controlled manner in all subsequent generations. Despite the development of sophisticated vectors capable of efficient transformation (see 7 for further discussion and references) so far genes have only been transferred using such vectors into a relatively narrow range of dicotyledonous plants. Direct gene transfer into protoplasts of monocotyledonous[8] and dicotyledonous[9,10] plants has also been achieved. As in animal cells, it has not yet been possible to insert the genes into a chosen region of the chromosome.

(c) Total and Partial Genome Transfer

In most cases the genes governing a given desirable trait have not been physically isolated; in others the trait is governed by several unidentified genes. It is not possible therefore to insert such genes into a vector as in (b). Classically such genes have been transferred by sexual crossing followed by back-crossing to breed out the unwanted portions of the genome. It is now possible to extend the ability of total genome transfer beyond the confines of sexual cross-compatibility using protoplast fusion. However, such wide-species fusions are only rarely likely to produce stable, fertile hybrids due to nuclear incompatibility. More likely applications of protoplast fusion may arise from partial genome transfer or from transfer of organelle genomes.

(i) Partial Genome Transfer. The aim here is to transfer randomly small pieces of the genome from a donor line (often a wide relative of a commercial crop) and select out the recipient line containing the desired trait. Such transfers may occur via fertilization with irradiated pollen[11,12] or by protoplast fusion[13] although to date the success has been limited. Much more detailed study is required on methods for controlling the fragmentation of the donor chromosomes and on determining the amounts of genetic material that can be transferred and subsequently inherited in a stable manner.

(ii) Organelle Genome Transfer. Besides the nuclear genome plants also contain two other genomes, those of the mitochondria and chloroplasts. Both of these encode important agronomic traits and situations can be envisaged where it is desired to transfer one or both of these genomes into a different nuclear background. Although this has been achieved in a limited way the process requires extending to a wider range of species. In general, if a plant cell contains two different mitochondrial genomes recombination between the genomes will occur at a significant frequency[14]. In contrast, recombination between chloroplast genomes is very rare and usually one of the two types present will come to dominate in plants regenerated from the original digenomic cell.

(d) Protoplast and Tissue Culture

It is obvious from the above that the ability to culture plant protoplasts and tissues is of vital importance in genetic manipulation. In particular the totipotency of plant cells (i.e. the ability to regenerate a whole plant from a single somatic cell) enables genes transferred into, or genetic changes induced in, a single cell to be incorporated into the germ line and thus inherited in subsequent generations. Protoplasts are thus seen as of potential importance as (a) recipients for transformation, (b) fusion partners for genome transfer and (c) an efficient means of obtaining a range of selected mutants. Tissue culture techniques are also relevant to transformation and mutant selection. The problems are that plant regeneration from protoplasts reliably and at high frequency is limited to a number of species, particularly in the Solanaceae[15]. The requirement is to extend the range of species that can be handled in this manner to include all the major world crops (e.g. cereals). A further problem is that tissue culture procedures appear to be mutagenic and the plants derived from culture show a wide range of variation (so called "somaclonal variation")[16,17]. There is also a need to improve techniques for fusion although several recent advances have been made[18].

IV. ACHIEVEMENTS

Much progress has been made in the last few years. I have selected three examples for discussion below.

(a) Analysis of Seed Storage Proteins. These genes for seed storage proteins were among the first to be cloned. Analysis of gene structure of the cereal seed prolamins has led to an understanding of how these proteins are constructed[19,20]. Essentially all prolamins appear to have multiple domains - one of which consists of blocks of repeated sequences of amino acids. In the case of the wheat proteins present in the protein portion of dough (termed gluten) are a set of high molecular weight proteins that are believed to confer elasticity, which is an important component in the ability to make bread from wheat four. Structural prediction based on amino acid sequences derived by recombinant DNA techniques[21] has led to the hypothesis that these proteins are largely in the form of stacked β turns (as distinct from α helix, β sheet or random coil) (see Figure 1) and these predictions

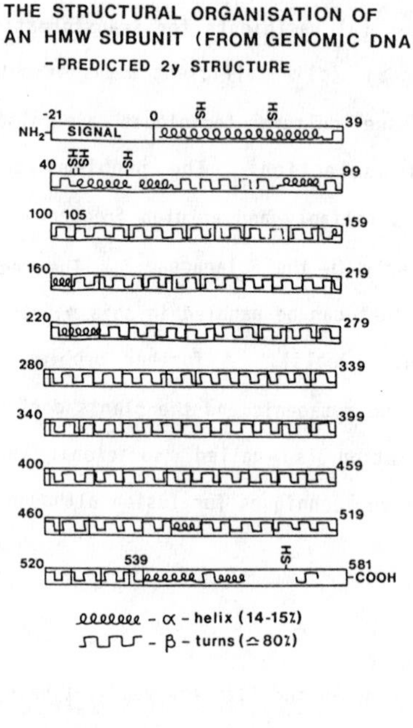

THE STRUCTURAL ORGANISATION OF
AN HMW SUBUNIT (FROM GENOMIC DNA)
-PREDICTED 2y STRUCTURE

NH_2 -21 SIGNAL 0 SH SH 39
40 SH SH SH 99
100 105 159
160 219
220 279
280 339
340 399
400 459
460 519
520 539 SH 581 -COOH

α - helix (14-15%)
β - turns (≃80%)

Figure 1. Predicted secondary structure of a sequence specified by a HMW glutenin subunit gene of wheat[21]. The predictions were carried out by Dr. A. S. Tatham as previously described [22,23] (previously unpublished work of Dr. Tatham).

have been confirmed by circular dichroism spectroscopy[22,23]. This structure has important parallels with that of elastin and in part explains the elasticity of these proteins. Information gained from these studies should prove of importance to breadmaking, a widespread and very ancient biotechnlogical process.

(b) Transfer of Genes under Expression Controls

The small subunit of ribulose bisphosphate carboxylase is encoded by genes in the nucleus, which are transcribed at high rates only in the light, is synthesized from mRNA on cytoplasmic ribosomes and is then transported to the chloropolast where it combines with the chloroplast-encoded large subunit to form the mature enzyme[24] (see Figure 2). It may therefore be expected

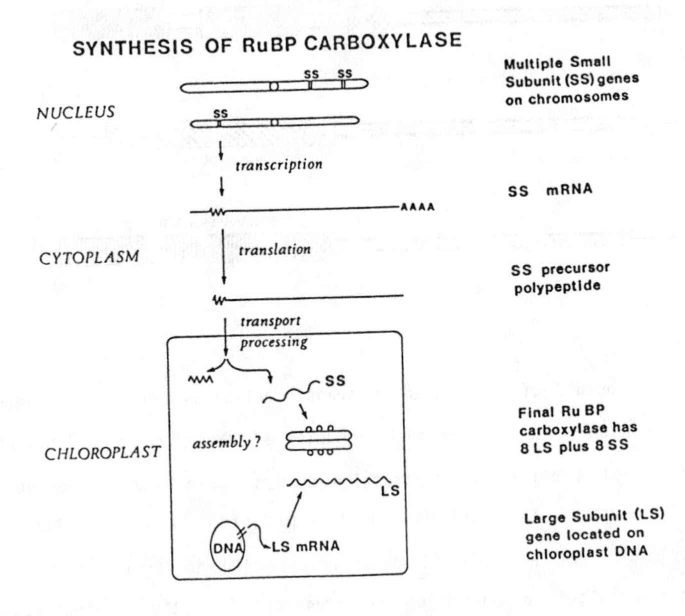

Figure 2. Diagram of the pathway of gene expression in the production of functional ribulose bisphosphate carboxylase (RUBISCO).

that sequences upstream to the final coding region would control light-dependent transcription and the transport of the pre-protein into the chloroplast (Figure 3a). Recent results have shown that when the small subunit gene from pea is transferred into petunia plants with about 1000 bases upstream from the coding sequence then this gene is transcribed in a light-dependent manner and the product properly processed and transported into the chloroplast[25,26]. The 5' sequences of this gene have also been spliced onto the coding sequence for the bacterial enzymes chloramphenicol

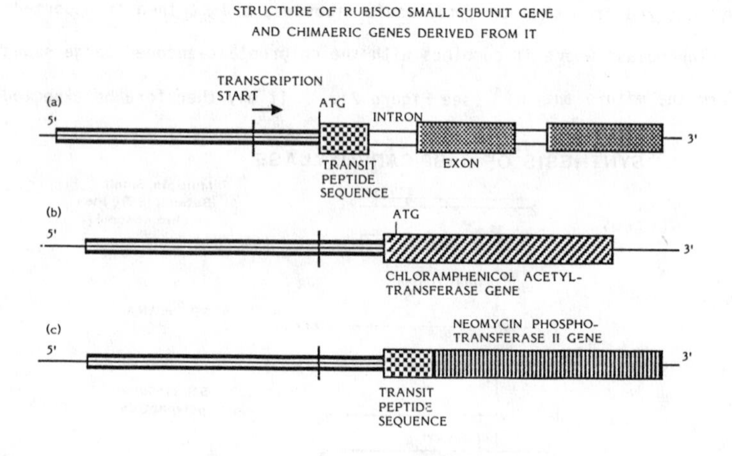

Figure 3. Structure of genes used in transformation experiments in higher plants: a) the pea gene for the small subunits of RUBISCO used to transform Petunia[25,26], b) a chimaeric construction used to transform tobacco cells and which resulted in light-dependent expression of chloramphenicol transacetylase[27], c) a chimaeric construction used to transform tobacco cells which resulted in expression and transport into the chloroplast of neomycin phosphotransferase[28]. In each case these genes were introduced into the plant cells using vectors based on the Agrobacterium tumefaciens Ti-plasmid.

transacetylase[27] (Figure 3b) and neomycin phosphotransferase II[28] (Figure 3c). In one experiment it was shown that the expression of the bacterial gene was then induced by light and in a second that the enzyme was processed and transported into the chloroplast. These results have thus identified sequences conferring environment-directed gene expression and enabling proteins to be targetted into plant organelles. Similar analysis of other regulated genes (*e.g.* for seed proteins) is in progress[7].

(c) Generation of Cybrids

Male sterility is an important trait that can be used in the production of hybrid seed. It is generally considered to be conferred by the mitochondrial genome. Resistance to the herbicide atrazine (also agronomically desirable in certain circumstances) is conferred by mutation occurring in the chloroplast genome. Although organelle encoded traits can be transferred by sexual means there are sometimes problems that are difficult to overcome. Recently it has been demonstrated that it is possible to transfer organelles containing these traits by protoplast fusion[29]. In one experiment two cytoplasmic characters belonging to different genera (*i.e.* male sterility encoded by the mitochondrial genome of *Raphinus sativus* and atrazine-resistance encoded by the chloroplast genome of *Brassica campestris*) were combined with the nuclear genome of a third species *Brassica napus* (see Figure 4). Although the plants produced still lack all the necessary attributes of desirable cultivars the results represent considerable progress.

CYBRID PRODUCTION BY PROTOPLAST FUSION

A.

GENOMES	PARENT 1	PARENT 2	SELECTED CYBRID
Nuclear	*B. napus*	*Brassica*	*Brassica*
Chloroplast	*B. napus*	*R. sativus*	*B. napus*
Mitochondrial	*B. napus*	*R. sativus*	*R. sativus*
PHENOTYPE			
Erucic Acid	Low	High	High
Normal Green Leaves	Yes	No	Yes
Atrazine Susceptible	Yes	Yes	Yes
Male Sterile	No	Yes	Yes
Well Developed Nectaries	Yes	No	Yes

B.

GENOMES			
Nuclear	*B. napus*	*Brassica*	*Brassica*
Chloroplast	*B. campestris*	*R. sativus*	*B. campestris*
Mitochondrial	*B. campestris*	*R. sativus*	*R. sativus*
PHENOTYPE			
Normal Green Leaves	Yes	No	Yes
Atrazine Susceptible	No	Yes	No
Male Sterile	No	Yes	Yes
Well Developed Nectaries	Yes	No	Yes

Figure 4. Characteristics of parent plants and selected cybrids produced after protoplast fusion. The information is drawn from the work of Pelletier et al. 1983[29,30].

V. THE FUTURE

The pace of research in this field is increasing enormously. I predict that in the next few years considerable progress to achieving the above aims will be met and in particular we will see:-

(a) A massive increase in useful analytical data enabling us to understand mechanisms controlling gene expression

(b) A great increase in the transfer of genetic information at all levels producing novel plants, some of which will have enhanced agronomic value

(c) The widening of protoplast and tissue culture technolgy to many species

(d) The development of a second generation of aims based on our increasing knowledge of plant metabolism and development derived from the use of the analytical powers of recombinant DNA technology.

REFERENCES

1 B. J. Miflin "The Genetic Manipulation of Plants and Its Application to Agriculture", Eds. P. J. Lea and G. R. Stewart, OUP, Oxford, 1985, p.295.

2 P. Dunsmuir, S. Smith and J. Bedbrook Nucleic Acid Res. 1984, 11, 4177.

3 J. V. Cullimore, C. Gebhardt and R. Saarelainen, B. J. Miflin, K. B. Idler and R. F. Barker J. Mol. Appl. Genet. 1984, 2, 589.

4 Z. Schwartz-Sommer and H. Saedler Oxford Surveys of Plant Molecular and Cell Biology 1985, 2, 353.

5 N. Fedoroff, D. B. Furtek and O. E. Nelson Proc. Natl. Acad. Sci. USA 1984, 81, 3825.

6 C. O'Reilly, N. S. Shepherd, A. Pereira, Z. Schwartz-Sommer, I. Bertram, D. S. Robertson, P. A. Peterson and H. Saedler EMBO J., 1985, 4, 877.

7 T. Hall Oxford Surveys of Plant Molecular and Cell Biology 1985, 2, 329.

8 H. Lorz, B. Baker and J. Schell Mol. Gen. Genet. 1985 in press.

9 F. H. Krens, L. Molendijk, G. J. Wullems, and R. A. Schilperoort Nature 1984, 311, 763.

10 J. Paszkowski, R. D. Shillito, M. Saul, V. Mandak, T. Hohn, and B. Hohn EMBO J. 1984, 3, 2717.

[11] P. D. S. Caligari, N. R. Ingram and J. L. Jinks Heredity 1981, 47, 17.

[12] J. W. Snape, B. B. Parker, E. Simpson, C. C. Ainsworth, P. I. Payne and C. N. Law Theor. Appl. Genet. 1983, 65, 103.

[13] P. P. Gupta, O. Schieder and M. Gupta Mol. Gen. Genet. 1984, 197, 30.

[14] M. Hanson Oxford Surveys and Plant Molecular and Cell Biology 1984, 1, 33.

[15] O. L. Gamborg and N. Dunn-Coleman "Advances in Gene Technology: Molecular Genetics of Plants and Animals" Eds K. Downey, R. W. Voellmy, F. Ahmad, J. Schultz, 1983, Miami Winter Symp. 20, p.101.

[16] P. J. Larkin and W. R. Scowcroft "Genetic Engineering of Plants" Eds T. Kosuge, C. P. Meredith and A. Hollaender, Plenum, New York, 1983, p.355.

[17] A. Karp and S. W. J. Bright Oxford Surveys of Plant Molecular and Cell Biology 1985 2, 199.

[18] M. J. Tempelaar and M. G. K. Jones Oxford Surveys of Plant Molecular and Cell Biology 1985 2, 347.

[19] P. Argos, K. Pedersen, M. D. Marks and B. A. Larkins J. Biol. Chem. 1982, 257 9984.

[20] M. Kreis, P. R. Shewry, B. G. Forde, J. Forde and B. J. Miflin Oxford Surveys of Plant Molecular and Cell Biology 1985 2, 253.

[21] J. Forde, J-M. Malpica, N. Halford, P. R. Shewry, O. Anderson, F. C. Greene and B. J. Miflin 1985, manuscript in preparation.

[22] A. S. Tatham, P. R. Shewry and B. J. Miflin FEBS Letts. 1984, 177, 205.

[23] A. S. Tatham, B. J. Miflin and P. R. Shewry Cereal Chem. 1985, in press.

[24] R. J. Ellis and A. Gatenby "The Genetic Manipulation of Plants and Its Application to Agriculture" Eds. P. J. Lea and G. R. Stewart, OUP, Oxford 1985, p.41.

25 G. Coruzzi, R. Broglie, C. Edwards, and N-H. Chua EMBO J. 1984 3, 1671.

26 R. Broglie, G. Coruzzi, R. T. Fraley, S. G. Rogers, R. B. Horsch, J. G. Niedermeyer, C. L. Fink, J. S. Flick and N-H. Chua Science 1984, 224, 838.

27 L. Herra-Estrella, G. Van Den Broeck, Maenhaut, M. Van Montagu, J. Schell, M. Timko and A. Cashmore Nature 1984, 310, 115.

28 G. Van den Broeck, M. P. Timko, A. P. Kausch, A. R. Cashmore, M. Van Montagu, L. Herrara-Estrella Nature 1985, 311, 358.

29 G. Pelletier, C. Primard, F. Vedel, P. Chetrit, R. Remy, P. Rousselle and M. Renard Mol. Gen. Genet. 1983 191, 244.

30 G. Pelletier, F. Vedel and G. Belliard Hereditas Suppl 1985 3 49-56.

7

The Application of Genetic Engineering to the Production of Pharmaceutical Compounds

By S. Harford

DEPARTMENT OF FERMENTATION DEVELOPMENT, GLAXOCHEM LIMITED, ULVERSTON, CUMBRIA LA12 9DR, U.K.

Biotechnology can be defined as the commercial application of engineering and technological principles to the life sciences. The history of biotechnology can be traced over many millenia and it has been described as the world's second oldest profession. Initially, the food and drinks industries were the main province of biotechnology with the manufacture of bread, beer, wine, cheese and many other fermentable products. Over more recent times the chemical and pharmaceutical industries have used biotechnological processes for the synthesis of many natural products, e.g. industrial alcohol, citric acid, a range of amino acids, antibiotics, vitamins, etc.

During the past decade there has been a major upsurge of interest in biotechnology. This has been brought about by the advent of recombinant DNA (rDNA) technology, otherwise known as gene cloning or genetic engineering. It was soon realised that the methods of genetic engineering greatly enhanced the potential of biotechnology providing the prospect for the development of many new products and bioprocesses. Biotechnology is now viewed by many as the final major development of the century, likely to have profound commercial and sociological effects in the 21st century. This has been documented in every conceivable publication, be it scientific journal, national magazine or daily newspaper such that the impact of rDNA technology on biotechnology is common knowledge.

Early economic forecasts predicted a bright future for the new industry and were sufficiently optimistic that they stimulated the growth of a new venture-capital based biotechnology sector. Over 100 new companies have now set up worldwide and indeed Nature has its own index of stocks for the larger of the biotechnology companies (1).

Although the impact of rDNA technology may eventually have a role in many commercial areas, Table 1, it is not surprising that the pharmaceutical industry has been the first to apply this methodology. Historically, the pharmaceutical sector has always worked with biological systems and has a great deal of experience in fermentation technology. Also the relatively low volume to high cost ratio of the fine chemicals produced can accommodate the great expense and investment required in developing a new technology of this type.

This article will concentrate on the pharmaceutical sector, observing the impact and application of rDNA technology. However, although the methods used will be dealt with in greater detail elsewhere in this volume it is nonetheless worthwhile to outline a few of the important developments before looking at their application in specific areas.

TABLE 1

Potential Areas of Importance for Genetic Engineering.

Area of Interest	Example of use
Food Industry	Production (single cell protein)
	Processing - Enzyme production (Rennin)
Agriculture - Crops	Crop improvements
	Crop resistance
	Nitrogen fixation capability
- Animals	Growth hormones
	Animal vaccines
Resources	Improvements in oil recovery
	Improvements in ore-leaching
	Ethanol production
	Pollution control
Chemicals	Amino-acids and vitamins
	Aliphatic/Aromatic synthesis

Gene Cloning

The revolution in biotechnology has been brought about by a relatively simple technological development. Genetic engineering methods enable the predetermined construction or dissection of a DNA segment, be it genomic, semi-synthetic or totally synthetic. DNA fabrication gives to the scientific community a new and extremely powerful tool for the study of biological systems.

The preparation of genomic libraries usuallyhas been accomplished by cloning large random, or semi-random, fragments of genomic DNA into an appropriate bacteriophage λ vector (2) or into a suitable cosmid (3). In this way total genomic representation can be achieved with a relatively small clone bank. This approach has been very successful in cloning and identifying specific genes within complex genomes. However, the presence of introns (4) in eucaryotic genomes has provided an additional problem in that the DNA sequence does not truly reflect the direct coding sequence for the protein product. In fact the mRNA transcribed from this DNA is processed to eliminate the intron regions and in this processed form encodes the protein sequence.

This property of mRNA processing has given rise to a second method of gene cloning *via* a semi-synthetic route. Total mRNA can be isolated from suitable eucaryotic cells and, using this as a template, a double stranded DNA copy can be synthesised *in vitro* using reverse transcriptase (5) and cloned into a suitable vector. A wide range of eucaryotic genes have been cloned by this cDNA route, *e.g.* human interleukin 2 (6), α - interferon (7) *etc*.

In some cases, where the entire amino-acid sequence of a protein is already known, it is possible to use total gene synthesis to give the desired clone. This approach involves designing a gene sequence that encodes the protein. Chemical synthesis of oligonucleotides provides the 'building blocks' for construction and these are ligated to produce the required gene sequence.

This method has been used for producing genes which encode small peptides (8) and also in some cases has been adopted for quite large genes (9, 10).

Gene Sequencing

DNA sequencing procedures (11, 12) have revolutionised protein sequence determination. Once a gene has been cloned its sequence can be analysed and any encoded protein sequence can be accurately predicted. Thus many of the problems of direct protein sequencing can be avoided. For example, many interesting eucaryotic proteins are only present in trace quantities which are inadequate for protein sequence determination. Also proteins must be in a highly purified state to ensure reliable sequence data. Hence, a predicted protein sequence can now be acheived within a matter of days instead of months or even years when employing the more conventional methods.

Site Directed Mutagenesis

A range of methods has been developed for the *in vitro* mutagenesis of cloned genes at high frequency. These fall into two types, either introducing random changes or 'true' site-specific modifications within the gene.

Random point-mutagenesis can be brought about by chemical treatment of DNA *in vitro*, *e.g.* by incubation with hydroxylamine (13) or by using bisulphite to modify single stranded regions created in DNA (14). Similarly, enzymic introduction of random point-mutants involves the creation of a small single stranded region within the DNA of the gene followed by deliberate mis-incorporation of nucleotides during the enzymatic re-synthesis of the second strand (15).

Specific site-directed mutagenesis, involving the modification of a pre-determined codon within a structural gene can be achieved using oligonucleotide mutagenesis (16).

These methods can be used to introduce mutations within a specific cloned gene at frequencies in excess of 10%. In addition to random changes it is also possible to replace specific amino-acid residues by others in order to study the effect on protein function. It is this technique which has given rise to the new field of protein engineering, enabling the design of proteins with altered structure-function relationships (17).

Gene Expression

This particular area has been researched extensively. The aim is to take the cloned gene encoding the desired protein and to express it at a high level. This often involves expressing the protein in a 'foreign organism' and the development of a highly sophisticated expression vector.

The expression hosts used include a number of bacteria, simple eucaryotes and mammalian cell lines. Most work has concentrated on using *Escherichia coli*, a natural choice given the understanding of the genetics and biochemistry of this organism. Bacillus species produce a number of excreted proteins and this property is potentially advantageous in aiding protein purification. Therefore, Bacillus have also been used as expression hosts (18). Yeasts, particularly *Saccharomyces cerevisiae*, being eucaryotic have the additional advantage that they may be able to carry out some of the post-translational modifications observed for many proteins from higher eucaryotes and have also been used for the production of proteins from cloned genes (19). Mammalian cell lines would be expected to correctly process mammalian proteins; however the technology for large scale growth of such cell lines is not as fully developed (20). It has also been proposed that Actinomycetes (21) and filamentous fungi may also have advantages as expression hosts.

A number of expression vectors have been developed for each of the hosts described above. Some of these vectors are highly versatile being suitable for the expression of many different proteins while others are designed to overcome specific expression problems. Most systems have been developed for expression in E. coli. Proteins may be expressed constitutively or following induction (22, 23). Some small peptides are more conveniently produced as a hybrid-molecule, fused to the end of a 'carrier protein. This fusion product can then be specifically cleaved to liberate the required peptide. The cleavage can be either chemical, e.g. using cyanogen bromide to cleave at methionine residues (24), or by employing a proteolytic enzyme having a known, and highly specific, recognition site for cleavage (25).

Expression vectors also exist for Bacillus (26), Yeasts (27) and mammalian systems (28, 29).

Having looked briefly at the available tools we can now turn to specific applications of the technology within the pharmaceutical sector. While most examples will outline on-going projects, some will demonstrate what has already been achieved by rDNA methods and others will show the way in which this technology may well play an important role in the future.

Fermentation Products

A wide range of pharmaceutical compounds are produced as secondary metabolites of microbial fermentations. Antibiotics are the most well known group of such compounds. The present day antibiotic production strains have been developed by several decades of strain selection following classical mutagenesis programmes (30). This highly successful, but empirical approach is likely to continue in the future. However, the more direct and selective application of rDNA methods will play an increasingly important role in generating more productive strains.

Changes which improve productivity are not always readily apparent but some possible ways of bringing about these changes are listed in Table 2. Increasing gene dosage or improving promoter strength may overcome potential rate-limiting steps in a biosynthetic pathway and can be achieved by standard gene cloning techniques. Various regulatory and control mechanisms can be eliminated or altered by gene cloning or site directed mutagenesis of regulator genes. It may be possible to improve certain properties of key enzymes by site directed mutagenesis. An alternative, but nonetheless complementary strategy, is to clone the enzymes for a catabolic pathway into the production organism thereby enabling it to utilise a cheaper growth substrate.

The genetics and biochemistry of secondary metabolite production is very poorly understood. It is clear that more fundamental research on the enzymes involved will be required before this approach can be gainfully employed.

TABLE 2

Applications of Genetic Engineering to Fermentation Products

1) Increase gene dosage of biosynthetic genes
2) Improve promoter activity
3) Eliminate feed-back control of synthesis
4) Modify biosynthetic genes to higher activity
    a) increase turnover number
    b) improve stability
5) Modify organism to use cheaper growth substrates
6) Modify physiology to improve fermentation, e.g. reduced viscosity

Another exciting possibility in this area is to use rDNA methods to modify microbes such that they produce novel secondary metabolites. This can be achieved using gene cloning to introduce new biosynthetic enzymes to create altered intermediates that will be incorporated into the structure of the final product. Alternatively, site-directed mutagenesis can be used to alter the specificity of existing enzymes so that they can utilise different substrate analogues as 'building blocks' for antibiotic structures. One such area already under investigation is the use of mutasynthesis and directed biosynthesis for the production of new antibiotics (31). It is clear that rDNA technology could play a major role in this development.

Protein Production

The direct application of rDNA technology to fermentation products, although of considerable economic significance, remains largely untested. In contrast, the large-scale production of commercially important proteins by rDNA methods has been the most successful use of the technology and the one which has received most publicity. Following a number of planned genetic manipulation steps a human protein, which normally may be only produced at a barely detectable level, can be synthesised in *E. coli* at a level in excess of 10% of total cellular protein corresponding to a yield of up to 1 gm/l. of fermentation broth. However, there are problems encountered when producing mammalian proteins in micro-organisms. Post-translational modifications (proteolytic cleavage or glycosylation) may be required or particular problems may be encountered during the self assembly of multi-subunit proteins or in obtaining correct secondary structure during the folding of the protein. A further problem is that many mammalian proteins form an insoluble complex when expressed at high level in micro-organisms (32). Even with such difficulties the understanding of mammalian protein expression in micro-organisms has been derived very rapidly and commercial interest in this aspect continues to be intense.

One of the first examples, and one that illustrates the success of this technology, was the production in *E. coli* of human growth hormone. This is clearly preferable to the conventional source of material which was protein derived from human cadavers. Similarly, human insulin is now produced commercially in *E. coli* under the trade name *Humulin*.

The number of human proteins being produced by genetically engineered *E. coli* is growing rapidly. Some may have anti-infective properties or may have a useful pharmacological activity. Indeed many of these proteins (Table 3) are the subject of clinical investigation to establish their therapeutic efficacy. Historically, very few proteins have been used clinically. One major reason for this has been the problem of producing large quantities of human proteins. Genetic engineering and microbial fermentation is removing this shortcoming for a growing number of proteins. Human proteins once available in only trace amounts can now be produced by the kilogram quantity from a modest fermentation capacity.

TABLE 3

Summary of Cloned Proteins and Their Uses

Potential Therapeutic Proteins	Envisaged Use
Insulin	Diabetes
Growth hormone Somatomedin C Growth hormone releasing factor	Growth promotion
Interferons (α,β,γ)	Anti-viral, anti-tumor, anti-inflammatory
Tumor necrosis factor Lymphotoxin	Anti-tumor
Interleukins - 1, 2 and 3 B-cell growth factors Macrophage activating factor Colony stimulating factors	Immune disorders Anti-tumor
α -1 Anti-trypsin	Emphysema
Serum Albumin	Plasma supplement
Factor VIII Factor IX	Haemophilia
Urokinase Tissue plasminogen activator Hirudin	Anticoagulant
Calcitonin	Osteomalacia
Epidermal growth factor	Wound healing
Erythopoietin	Anaemia
Urogastrone	Anti-ulcerative

Study of New Drug Targets

Many potential drug targets have been identified. These include key metabolic enzymes, growth factors, hormones, transmitter substances, oncogene products, neuropeptides, various receptor proteins, etc. The power of rDNA technology can be directed at these targets in order to characterise them more fully. DNA sequencing analysis can be used to predict the amino-acid sequence of cloned 'target genes' and the proteins can be expressed in sufficient quantity to provide material for X-ray crystallographic studies. The effect of changes brought about by site directed mutagenesis can be demonstrated directly in structure/function terms. Such knowledge is essential as an aid to computer assisted drug-design programmes in the future. The approach used may well follow the current work on the structure of renin (33) from which a number of anti-hypertensive agents have been designed.

Vaccine Development

This is another area where rDNA methods have proved successful. In the past, vaccine development used empirical methods to derive attenuated or killed vaccines to increase the safety of the product. Recombinant methods enable the researcher to dissect the gene for the active immunogen from the host organism (pathogen) and to introduce it into a more convenient and benign system for high level expression.

Such a system offers many advantages both in product safety and in productivity although in many cases product efficacy is still under review. A number of vaccine targets are currently under investigation and some of these are listed in Table 4. Clinical evaluation is underway currently for a number of these vaccines.

The hepatitis B vaccine may also prove to be in effect an anti-tumor vaccine in so far as hepatitis B infection is closely linked to hepatocarcinoma. Similarly, the development of a human papilloma virus vaccine could be effective against cancer of the cervix and vulva. Indeed, in the future it may be possible to design specific tumor-antigen based vaccines as a general approach.

However, it must be remembered that activation of the immune system is highly complex and the use of recombinant vaccines introduces many new problems. These same problems will be encountered when using synthetic peptides as vaccine products(34).

TABLE 4

Vaccine Development Using Genetic Engineering

Some Vaccines Under Investigation

Hepatitis B

Herpes

Malaria

Pertussis

Dental Caries

Rabies

Cholera

The above outline of applications of rDNA within the pharmaceutical industry follows progress from the early successes in protein and vaccine production to some future potential uses in the fermentation sector and in drug design. The overall importance of rDNA methods to the industry is difficult to assess as its research impact may turn out to be of far greater value than the immediate returns from product development and process improvement. Many products such as insulin, growth hormone, α-interferon and hepatitis B vaccine have been through extensive clinical evaluation and are already, or soon will be, commercial products.

It is clear that this technology will have a key role in research applications in the pharmaceutical industry. However, its commercial success will be judged by the profitability of the products it helps to bring to the market place over the next decade. The technology also brings its own problems in terms of product registration (35) and patent deliberations (36). However, for researchers actively engaged in the area, and particularly those in the pharmaceutical sector, this period will be both exciting and crucial for the overall growth and prosperity of the industry.

References

1. Anonymous, Nature, 1984, 310, 94.
2. O. Smithies, A. E. Blechl, K. Denniston-Thompson, N. Newall, J. E. Richards, J. L. Slightom, B. W. Tucker and F. R. Blattner, Science, 1978, 202, 1284.
3. D. Ish-Horowicz and J. F. Burke, Nucleic Acids Res., 1981, 9, 2989.
4. W. Gilbert, Nature, 1978, 271, 501.
5. J. G. Williams, "Genetic Engineering", Academic Press, New York, 1981, Vol. 1, p.2.
6. T. Tanaguchi, H. Matsui, T. Fujita, C. Takaoka, N. Kashima, R. Yoshimoto and J. Hamuro, Nature, 1983, 302, 305.
7. D. V. Goeddel, D. W. Leung, T. J. Dull, M. Gross, R. M. Lawn, R. McCandliss, P. Seeburg, A. Ullrich, E. Yelverton and P. W. Gray, Nature, 1981, 290, 20.
8. K. Itakura, T. Hirose, R. Crea, A. D. Riggs, H. L. Heyneker, F. Bolivar, H. W. Boyer, Science, 1977, 198, 1056.
9. M. D. Edge, A. R. Greene, G. R. Heathcliffe, P. A. Meacock, W. Schuch, D. B. Scanlon, T. C. Atkinson, C. R. Newton and A. F. Markham, Nature, 1981, 292, 756.
10. J. Smith, E. Cook, I. Fotheringham, S. Pheby, R. Derbyshire, M. A. W. Eaton, M. Doel, D. M. J. Lilley, J. F. Pardon, T. Patel, H. Lewis and L. D. Bell, Nucleic Acids Res., 1982, 10, 4467.
11. F. Sanger, S. Nicklin and A. R. Coulson, Proc. Natl. Acad. Sci. USA, 1977, 74, 5463.

12. A. M. Maxam and W. Gilbert, Proc. Natl. Acad. Sci. USA, 1977, 74, 560.

13. G. O. Humphreys, G. A. Willshaw, H. R. Smith and E. S. Anderson, Molec. gen. Genet, 1976, 145, 101.

14. P. E. Giza, D. M. Schmit and B. L. Murr, Gene, 1981, 15, 331.

15. D. Shortle, P. Grisafi, S. J. Benkovic and D. Botstein, Proc. Natl. Acad. Sci. USA, 1982, 79, 1588.

16. M. J. Zoller and M. Smith, Nucleic Acids Res., 1982, 10, 6487.

17. E. T. Kaiser, Nature, 1985, 313, 630.

18. S. D. Ehrlich, S. Jupp, B. Niaudet and A. Goze, "Genetic Engineering", 1978, Elsevier/North-Holland Biomedical Press.

19. R. Derynck, A. Singh and D. V. Goeddel, Nucleic Acids Res., 1983, 11, 1819.

20. J. Feder and W. R. Tolbert, Scientific American, 1983, 248, 24.

21. K. F. Chater, B. A. Hopwood, T. Kieser and C. J. Thompson, Curr. Topics Microbiol. Immunol., 1982, 96, 69.

22. E. Remaut, P. Stamssens and W. Fiers, Gene, 1981, 15, 81.

23. E. Amann, J. Brosius and M. Ptashne, Gene, 1983, 25, 167.

24. R. Wetzel, D. G. Kleid, R. Crea, H. L. Heyneker, D. G. Yansura, T. Hirose, A. Krasyewski, A. D. Riggs, K. Itakura and D. V. Goeddel, Gene, 1981, 16, 63.

25. K. Nagai and H. C. Thøgersen, Nature, 1984, 309, 810.

26. D. G. Yansura and D. J. Henner, Proc. Natl. Acad. Sci. USA, 1984, 81, 439.

27. R. A. Kramer, T. M. DeChiara, M. D. Schaber and S. Hilliker, Proc. Natl. Acad. Sci. USA, 1984, 81, 367.

28. Y. Gluyman, "Eucaryotic Viral Vectors", 1982, Cold Spring Harbor Laboratory, New York.

29. B. H. Howard, Trends Biochem. Sci., 1983, 8, 209.

30. Y. Aharonowitz and G. Cohen, Scientific American, 1981, 245, 106.

31. C. A. Claridge, "Basic Life Sciences", 1983, Plenum Press, New York, Vol. 25, p. 231.

32. D. C. Williams, R. M. Van Frank, W. L. Muth and J. P. Burnett, Science, 1982, 215, 687.

33. T. Blundell, B. L. Sibanda and L. Pearl, Nature, 1983, 304, 273.

34. P. Newmark, Nature, 1983, 305, 9.

35. D. A. Espeseth, G. P. Shibley, P. L. Joseph, R. A. Van Deusen and C. A. Whetstone, "Developments in Biological Standardization", 1985, S. Karger, Basel, Vol. 59, p. 167.

36. R. S. Crespi, Trends Biochem. Sci., 1982, 7, 423.

8

Clinical Applications of Molecular Biology

By C. G. P. Mathew

M.R.C. RESEARCH UNIT FOR MOLECULAR AND CELLULAR CARDIOLOGY, UNIVERSITY OF STELLENBOSCH MEDICAL SCHOOL, STELLENBOSCH, SOUTH AFRICA

Since its infancy the study of molecular biology has been regarded as a very academic pursuit of no relevance to clinical medicine. However this situation has changed dramatically during the past six or seven years, to the extent that the pages of the leading medical journals and the popular press are filled with the latest exploits of the 'Gene Doctors'. This change has been brought about by the extraordinary power of recombinant DNA technology, which has had a rapid and significant impact on medical research. This chapter will highlight some of the achievements and the potential of the new technology in the prevention, diagnosis and treatment of human disease.

MOLECULAR PATHOLOGY OF INHERITED DISEASE AND CANCER

β thalassaemia is an inherited disorder in which affected individuals are unable to synthesize the β chains of haemoglobin. Patients develop a severe anaemia which usually leads to death before the age of twenty, in spite of repeated blood transfusions. Molecular cloning of the normal β globin gene and determination of its nucleotide sequence led to cloning of these genes from patients with β thalassaemia. In most cases the defect was found to be a mutation of a single base within the 1600 nucleotide sequence of the gene. At least 22 different mutations have now been reported[1]. Some create premature translational stop signals in the mRNA (CAG → UAG), others alter signal sequences involved in processing of the primary RNA transcript so that no mature mRNA is produced, and several in the promoter regions of the gene affect its transcription. This precise definition of the defect at the molecular level has not only extended our understanding of gene regulation, but has led to the development of strategies for prenatal diagnosis of these disorders (see following section).

Although β thalassaemia was the first inherited disorder to be studied in such detail, the technology can be applied to any inherited disorder for which the identity of the affected gene is known. Over 400 human genes have been cloned[2], and the molecular basis of many genetic diseases is now being established.

Perhaps the most striking success of the new technology has been the discovery of oncogenes, the genes which cause cancer[3]. DNA isolated from tumour cells was found to transform mouse fibroblast cells in culture. The transforming principle was purified and shown to be a single gene, which had a normal cellular counterpart (a proto-oncogene). Comparison of the sequence of the normal gene with that of its oncogene isolated from a bladder carcinoma showed that they differed by only 1 nucleotide out of 5000. Thus a point mutation, which causes a single amino acid substitution in the protein for which it codes, is sufficient to activate the oncogene. The role of the normal oncogene proteins in cellular metabolism is now being studied intensively, and some have been shown to be involved in the control of cell division and growth. The elucidation of the molecular events involved in tumour evolution is much more likely to lead to a rational strategy for cancer therapy than the empirical approaches which have had to be adopted in the past.

DIAGNOSIS OF INHERITED DISEASE

The major clinical application of molecular biology thus far has been in the diagnosis of inherited disease. DNA isolated from the patient's white blood cells is digested with sequence-specific restriction endonucleases, separated by electrophoresis in agarose gels, and transferred to a nitrocellulose filter. The filter is then incubated with a radioactively labelled copy of a specific gene (the 'probe'), which anneals only to its complementary sequences on the filter. DNA fragments containing the gene are detected by autoradiography. This technique, usually referred to as the Southern blot[4], allows one to analyse the structure of any gene for which a probe is available. It is the basis of several different approaches for the diagnosis of inherited disease.

(i) Restriction enzyme analysis

Sickle cell anaemia is an inherited disorder which results from an amino acid substitution in the sixth residue of the β chain of haemoglobin ($\beta^{6\ glu \rightarrow val}$). The DNA sequence is mutated from CCT-GAG-GAG to CCT-GTG-GAG. The restriction enzyme Mst II cleaves DNA at the sequence CCT NA GG (where N = any of the 4 bases). Thus the sickle mutation results in the loss of a cleavage site for Mst II in this region of the gene. The larger restriction fragment produced from the sickle gene can readily be detected by probing Southern blots with the β globin gene (figure 1).

Figure 1

DNA was digested with Mst II, electrophoresed on a 1% agarose gel, blotted onto nitrocellulose, and incubated with a ^{32}P- labelled β globin probe. The normal (β^A) globin gene is located on a 1.1 kilobase (Kb) fragment, and the sickle (β^S) gene on a 1.3 Kb fragment. Lane a = Hb AA, b-c = Hb SS, d = Hb AS.

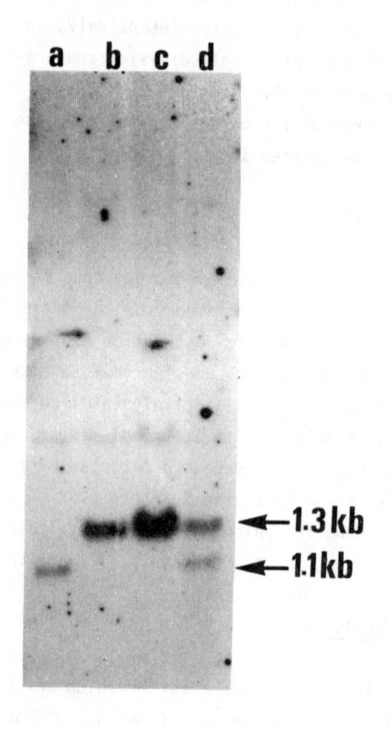

(ii) Synthetic oligonucleotides

Many of the point mutations which cause genetic disease do not fall within the recognition sequence of a restriction enzyme. In such cases, synthetic oligonucleotides can be used as probes to detect single base mismatches[5]. Two 19-mer olignucleotide probes are synthesized with sequences corresponding to the normal and the mutated gene. The probes are then labelled with ^{32}P, and incubated with filters containing the patient's DNA. Under carefully controlled conditions, the probe will hybridize only to its homologous sequence. Thus if a positive 'signal' is detected with the mutant probe, the abnormal gene is present in the patient's DNA. This approach has been used for the diagnosis of sickle cell anaemia, β thalassaemia and α_1 - antitrypsin deficiency[5]. Very recently it has been used to identify activated oncogenes in tumour DNA[6].

(iii) Restriction enzyme polymorphisms

The approaches described above are only feasible if the precise mutation which causes the defect is known. If it is not, an indirect approach can be used which exploits neutral mutations within or close to the gene which alter the cleavage pattern of the gene by a particular restriction enzyme. Such mutations, referred to as restriction enzyme polymorphisms, do not affect the expression of the gene and occur at a relatively high frequency in the normal population. They can be used as genetic markers to follow the inheritance of abnormal genes in families. For example, the enzyme Rsa 1 cleaves a human collagen gene to produce a 2.1 Kb or a 2.9 Kb fragment[7]. We have studied a family with an inherited disorder of collagen, osteogenesis imperfecta, in which the affected grandmother had both the 2.1 and 2.9 Kb fragment, whereas her husband had only the 2.1 Kb fragment[7]. We found that all affected individuals in further generations of the family had inherited the 2.9 Kb fragment, whereas unaffected members had not. Thus the abnormal collagen gene in this family was associated with an Rsa 1 polymorphism which resulted in the production of a larger restriction fragment. The mutation which eliminated a cleavage site for Rsa 1 did not cause the disease, because it occurred in about 50% of the normal population.

The value of these techniques for the postnatal diagnosis of genetic disease is considerable. They are also useful for the detection of unaffected or silent carriers of a recessively inherited disorder, which is only manifest in individuals who have a double 'dose' of the abnormal gene. However the great power of the approach derives from the fact that it can be used for the prenatal diagnosis of genetic disease. Fetal DNA can be obtained from chorionic villus biopsy at the 8th week of pregnancy[8], or from cultures of amniotic fluid cells, and analysed for the presence of a particular gene defect. Since all somatic cells contain a full complement of genes, it should be possible to diagnose any disorder for which a DNA probe is available. DNA analysis is already the method of choice for the prenatal diagnosis of sickle cell anaemia and thalassaemia[1], and is rapidly being applied to many other types of genetic disease.

GENE THERAPY

The identification of the affected gene in a number of inherited disorders and the availability of cloned copies of their normal counterparts has created the possibility of correcting a genetic defect by supplying the patient with a normal functional gene[9]. Perhaps the most likely candidate for this approach is thalassaemia. The disease is well understood, the normal gene is available, and the tissue affected by the disease (bone marrow and blood) can be treated *in vitro* and transplanted back into a patient. Current research is directed towards providing treated red blood cells with a selective advantage over untreated cells so that they predominate *in vivo*, and to introducing the normal β globin gene into the cells in such a way that it is efficiently expressed.

An even more radical approach is embryo therapy, in which the gene is injected into the early embryo. For example, 'supermice' have been produced by injecting rat growth hormone gene into mouse embryos[10]. This procedure raises serious ethical problems because it would affect not only the individual who was treated, but also his or her offspring[9].

PRODUCTION OF DRUGS

Since human genes can be inserted into bacteria and expressed, it is now possible to engineer new bacterial strains which produce clinically important drugs in quantity. A notable example of this is the anti-viral and anti-cancer drug interferon, which occurs in trace amounts in white blood cells. An *E. coli* strain which produces large quantities of interferon has been made[11] and the drug is now undergoing extensive clinical trials which were previously not possible because of the limited supplies available. Other drugs produced in this way include human insulin, growth hormone and somatostatin.

PRODUCTION OF VACCINES

Vaccinia virus, which was used originally for the eradication of smallpox, is now being genetically engineered to create new live vaccines for use against many other viral infections[12]. The strategy employed is to insert the foreign DNA (for example, the gene coding for the hepatitis B virus surface antigen HBsAg) into a plasmid which contains a vaccinia virus transcriptional promoter upstream of the insertion point. The plasmid also has short vaccinia virus DNA sequences flanking the cloning site which are required for insertion of the foreign DNA into the viral genome. The recombinant virus can then be selected for, purified, and used for vaccination. In order to test this approach chimpanzees were given a single vaccination of a recombinant virus which expressed HBsAg, and challenged after 14 weeks with an intravenous injection of hepatitis B virus[12]. The vaccinated animals did not develop hepatitis.

Vaccinia virus recombinants have now been designed to protect against other viral infections such as influenza and herpes simplex. Attempts are also being made to produce a multiple vaccine by inserting up 20 different foreign genes into the vaccinia genome. Other advantages of the system are that it is cheap, stable, and simple to administer. An alternative strategy, which avoids the complications which occasionally occur from infection with a live virus, involves the production of the viral antigens in bacteria. The purified peptides can then be injected into the individual. However the peptides are poor antigens compared to viral particles, so the level of immunisation is low.

FUTURE PROSPECTS

The discovery and characterisation of oncogenes should lead to rapid progress in the diagnosis and treatment of cancer, and to identification of individuals or families at risk for the disease. Prenatal diagnosis of a large number of severe inherited diseases is now becoming possible, as the number of cloned human DNA sequences is increasing exponentially. Finally, the production of vaccines by recombinant DNA technology should result in the eradication of viral diseases such as hepatitis, and possibly even AIDS.

REFERENCES

1. S.H. Orkin and H.H. Kazazian, Ann. Rev. Genet., 1984, 18, 131.
2. J. Schmidtke and D.N. Cooper, Human. Genet., 1984, 67, 111.
3. R.A. Weinberg, Sci. Amer., 1983, 249, 102.
4. E.M. Southern, J. Mol. Biol., 1975, 98, 503.
5. K. Itakura, J.J. Rossi and R.B. Wallace, Ann. Rev. Biochem., 1984, 53, 323.
6. J.L. Bos, M. Verlaan-de Vries, A.M. Jansen, G.H. Veeneman, J.H. Boom and A.J. van der Eb, Nucl. Acids Res., 1984, 12, 9155.
7. A.F. Grobler-Rabie, G. Wallis, D.K.Brebner, P. Beighton, A.J. Bester and C.G. Mathew, EMBO J., 1985, 4(7), in press.
8. R. Williamson, J. Eskdale, D.V. Coleman, M. Niazi and B.M. Modell, Lancet, 1981, ii, 1125.
9. R. Williamson, Nature, 1982, 298, 416.
10. R.D. Palmiter, R.L. Brinster, R.E. Hammer et al., Nature, 1982, 300, 611.
11. S. Nagata, H. Taira, A. Hall et al., Nature, 1980, 284, 316.
12. G.L. Smith and B. Moss, Bio Essays, 1984, 1, 120.

9

Generation and Use of cDNA Clones for Studying Gene Expression

By M. J. Maunders, A. Slater, and D. Grierson

DEPARTMENT OF PHYSIOLOGY AND ENVIRONMENTAL SCIENCE, UNIVERSITY OF NOTTINGHAM SCHOOL OF AGRICULTURE, SUTTON BONINGTON, LOUGHBOROUGH LE12 5RD, U.K.

Introduction

cDNA cloning involves the copying of mRNA molecules into complementary single-stranded DNA (copy DNA), using the enzyme reverse transcriptase. The cDNA is then converted to a double stranded molecule, using DNA polymerase, and this double-stranded DNA is then inserted into a cloning vector (such as a plasmid, virus or cosmid), using various enzymes that cut and join DNA molecules. The cloning vector is then introduced into a suitable host cell, such as *Escherichia coli*, where it is replicated. By suitable handling of the bacterial cultures, individual cells can be isolated, each harbouring a vector with a different cDNA. During subsequent growth of the cells, many copies of the vector plus cDNA can be generated. The culture of individual cells thus produces clones, each carrying different cDNA molecules.

The importance of cDNA cloning is that it is the most effective method of purification of an mRNA sequence. The significance of the technique becomes obvious when it is considered that most higher eukaryotic cells contain 10,000-40,000 different mRNA sequences. Although cDNA cloning procedures were originally developed using mRNAs present in high concentration (abundance) in specialised cell types (*e.g.* globin mRNA, ovalbumin mRNA) it is theoretically possible to clone any mRNA sequence, even if present in only a few copies per cell, provided that a sufficient number of clones are generated to ensure that every mRNA sequence is represented in the cDNA clone population (a cDNA clone "library"). In practice, the major limitation lies in the detection of low abundance mRNA clones rather than in their generation. Once cDNA clones for specific mRNAs are identified, they can be used for a variety of purposes including DNA sequencing (which also predicts the amino acid sequences of the corresponding protein), the identification of the corresponding genomic sequences (leading to studies of gene control regions) and in hybridization assays to investigate changes in transcription of the genes during normal development or in mutants. Even if individual cDNA clones cannot be identified the procedure can still yield valuable information about gene numbers and control of gene expression.

Isolating the mRNA template. The twin aims of any mRNA isolation procedure will be to obtain an mRNA sample which is both undegraded and free from contaminants which might affect cDNA synthesis. The strategy adopted will depend upon whether a complete cDNA library or a sample enriched in one particular mRNA species is required. A standard form of extraction involves deproteinisation of whole cells, or a cytoplasmic fraction, with phenol and detergents such as sodium dodecyl sulphate. Alternatively, RNA can be extracted from tissues rich in ribonucleases with chaotropic agents (guanidinium hydrochloride or thiocyanate)[1]. RNA can be further separated from DNA, carbohydrate or protein contaminants by virtue of its relative insolubility in high salt solutions (*e.g.* 3 M sodium acetate) or its high buoyant density, by centrifugation through CsCl[2].

Since mRNA constitutes approximately 1% of most eukaryotic cytoplasmic RNA, mRNAs which carry a poly(A) tail at the 3' end can be further purified by affinity chromatography on oligo(dT) cellulose[3] or poly(U) sepharose[4].

It is also possible to enrich a mRNA population for specific mRNA sequences. The accumulation of specific mRNAs may be inducible *in vivo* or the mRNA in question may be localised in a particular cell fraction (*e.g.* on membrane bound polysomes) which can be purified. In some cases, mRNAs have been purified by immunoprecipitation of polysomes, utilising the ability of antibodies raised against a purified protein to recognise the nascent polypeptide chain being synthesised on ribosomes translating the corresponding mRNA[5]. More generally, it is possible to enrich for a particular mRNA by size fractionation of the total RNA extract by sucrose gradient centrifugation or gel electrophoresis[6]. This, of course, depends on the ability to recognise in which size fraction the mRNA of choice appears.

mRNA characterisation by in vitro translation. Eukaryotic mRNA characterisation is based on the ability of cell extracts (for example, rabbit reticulocyte lysate or wheat germ extract) to translate heterologous mRNAs *in vitro*. Translation of endogenous mRNAs can be prevented by prior treatment of the extract with micrococcal nuclease[7]. In the presence of added radioactively labelled amino acids, labelled proteins are synthesised and can be separated by SDS-polyacrylamide gel electrophoresis and visualised by autoradiography or fluorography using X-ray film. Figure 1 shows a fluorograph of a gel on which the ^{35}S-methionine-labelled translation products of tomato fruit mRNAs have been separated. The fruit mRNA population generates a number of discrete bands of different size and intensity, each band being the translation product of an individual mRNA species. The intensity of each band is a rough indication of the mRNA abundance, assuming that each mRNA is translated with the same efficiency and encodes a similar proportion of methionine residues. SDS-polyacrylamide gels can resolve up to 50 bands, which will correspond to the most abundant mRNAs. Even so, it is obvious in Figure 1 that a number of changes in the tomato fruit mRNA population which occur during ripening can be observed using this technique. Much higher resolution of up to 1,000 proteins can be obtained by 2-dimensional electrophoresis[8].

mRNA identification by immunoprecipitation of translation products. Although it is possible to determine the molecular weight of a translation product by SDS-polyacrylamide gel electrophoresis, using molecular weight marker proteins (Figure 1), comparison of size with a native protein may not be sufficient to identify the mRNA. Many proteins undergo proteolytic processing and postranslational modifications such as glycosylation, *in vivo*, which do not take place in translation systems *in vitro*. However, antibodies raised against a purified protein will recognise regions of the precursor polypeptide synthesised *in vitro*, and it is possible to specifically precipitate the precursor from the total translation mix and determine its size by gel electrophoresis (Figure 1). In this way, the mRNA coding for any protein of interest can be positively identified, provided that antibodies to the protein are available.

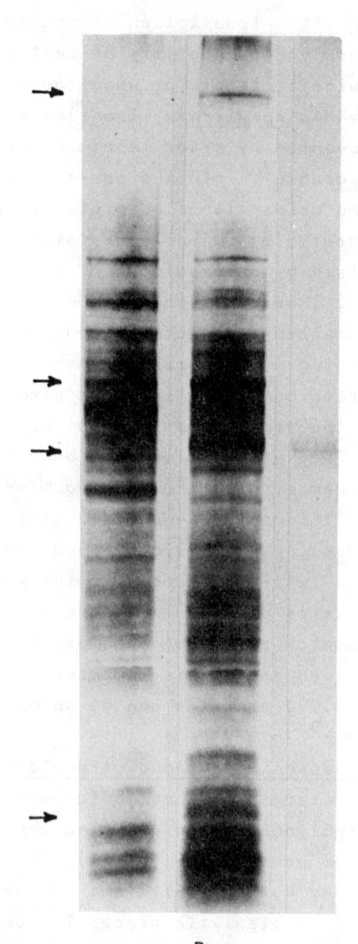

Figure 1 In vitro translation and immunoprecipitation

In this example, cytoplasmic RNA from green and red tomato fruit was translated in a rabbit reticulocyte lysate containing (^{35}S) methionine. The translation products were separated by SDS-polyacrylamide gel electrophoresis and detected by fluorography. The arrows indicate some mRNAs which increase during fruit ripening. The sample labelled "PG immune precipitate" was obtained by treating total red RNA translation products with antibody raised against purified polygalacturonase (PG). The immunoprecipitated translation product corresponds in size to one of the mRNA translation products which increase during ripening.

cDNA cloning

First strand synthesis Given as starting material a total or partially enriched poly(A)$^+$ mRNA population, cDNA copies of each molecule can be generated by reverse transcriptase (RNA-dependent DNA polymerase from avian myeloblastosis virus[9,10]). This enzyme will copy an RNA template by polymerising deoxyribonucleotide triphosphates (dNTPs) in a 5' to 3' direction, and requires a primer. In the case of poly(A)$^+$ mRNA, the poly(A) tail provides a very convenient site for annealing oligo (dT) to prime cDNA synthesis (Figure 2).

In theory, a cDNA molecule primed at the 3' end of the mRNA would result in the entire length of the mRNA molecule being copied. In practice, it is not always easy to synthesise full length cDNA, either due to ribonuclease contamination of the reverse transcriptase[11] or nucleolytic activities which are an integral part of the enzyme[12]. Secondary structure in some mRNA templates also inhibits full length synthesis but can be avoided by denaturing the mRNA just before the reaction. It should be noted that for many purposes a full length cDNA copy may not be essential.

The efficiency of 1st strand synthesis can be monitored by measuring the incorporation of a labelled dNTP into cDNA and the size range of labelled product can also then be checked by gel electrophoresis and autoradiography (Figure 3).

2nd strand synthesis. After first strand synthesis, the mRNA template is removed by alkali treatment and the second strand is synthesised using either the Klenow fragment of *E. coli* DNA polymerase or reverse transcriptase. Although both these enzymes require a primer, the 3' end of the first strand can form a transient hairpin structure which acts as a self-primer for second strand synthesis[13]. Second strand synthesis can also be monitored by size analysis of the labelled DNA on denaturing gels, since, due to the presence of the hair pin loop, a full length double stranded molecule should be twice as long as a single strand (Figure 3).

Removal of hairpin loop. At this stage the hairpin loop is removed by treatment with S_1 nuclease, a nuclease with a strong preference for single stranded nucleic acids. This reaction must be carefully monitored to ensure maximum cutting of the

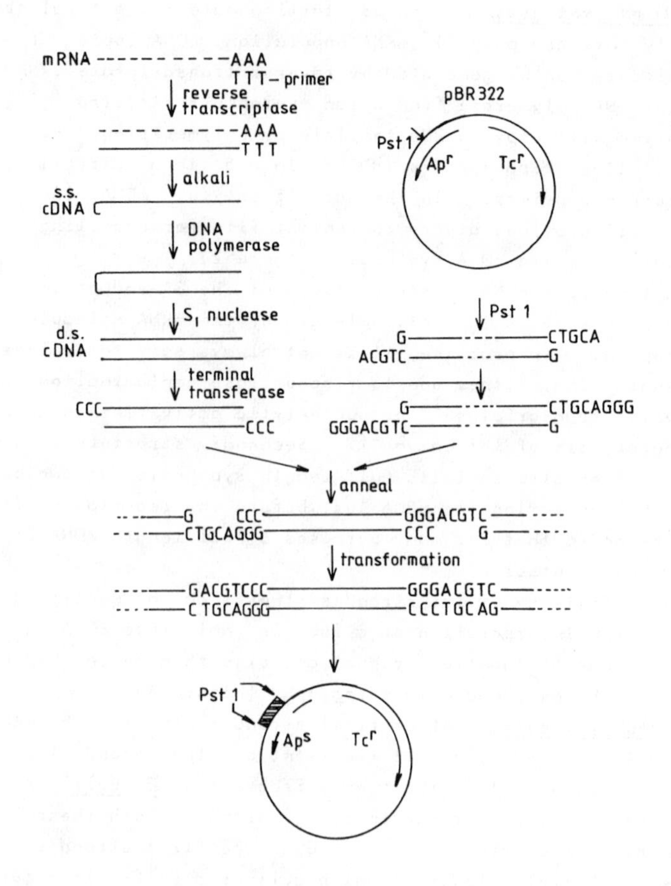

Figure 2 cDNA synthesis and cloning

The diagram outlines a scheme for synthesis of double stranded cDNA and insertion into a plasmid by homopolymer tailing. Reverse transcriptase is used to synthesise cDNA on the poly(A)$^+$ mRNA template, using oligo(dT) as a primer. After removal of the RNA, the second cDNA strand is synthesised utilising the transient hairpin loop which forms at the 3' end of the first strand as a primer. This loop is then cleaved and oligo(dC) tails are added to the 3' ends of the cDNA. The tailed cDNA can be inserted into the plasmid vector pBR322, prepared by cutting with Pst-1 and tailing the cut 3' ends with (dG) residues. The recombinant plasmid is used to transform a suitable E. coli host strain. The host cell repairs the gaps in the recombinant DNA, forming a Pst-1 site on each side of the cDNA insert.

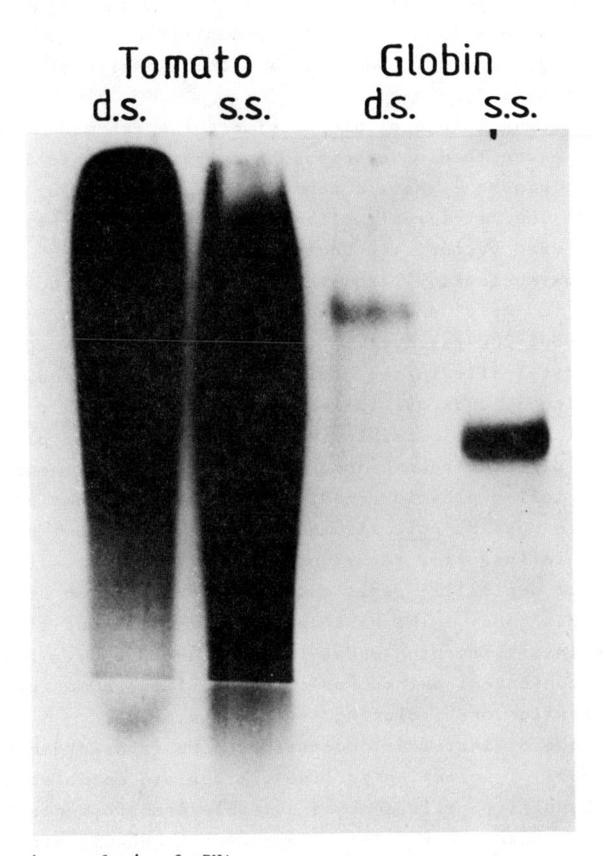

Figure 3 <u>Size analysis of cDNA</u>

cDNA was synthesised as described in Figure 2 in the presence of $\alpha(^{32}P)$ dCTP using tomato poly(A)$^+$ mRNA or purified globin mRNA as a template. The size of single stranded (s.s.) and double stranded (d.s.) cDNA before S_1 nuclease treatment was determined by agarose gel electrophoresis under denaturing conditions and the labelled cDNA was detected by autoradiography. The globin single stranded cDNA consists mainly of a single band with very little smaller material indicating efficient full length transcription. The double stranded cDNA is twice the length of the single strand under denaturing conditions, due to the presence of the hairpin loop. The plant poly(A)$^+$ mRNA consists of a population of many different mRNAs (see Figure 1) and generates a wide range of cDNA molecules, some many times larger than globin mRNA. Synthesis of double stranded cDNA increases the size of the cDNA population, consistent with each molecule doubling in size.

hairpin loops whilst minimising excessive degradation of the double stranded cDNA. It ay be desirable at this stage to perform a size selection p ocedure by centrifugation or column chromatography such that cDNAs below a certain size are discarded (Figure 4).

Insertion of cDNA into a cloning vector The trimmed double stranded cDNA can then be processed for insertion into the vector of choice. Figure 2 shows a common procedure used for insertion of cDNA into the Pst-1 restriction endonuclease recognition site of a widely used vector, the bacterial plasmid pBR322[15]. Terminal deoxynucleotidyl transferase is used to add dC residues on to the 3' ends of the cDNA strands, to an optimum length of about 20C residues. The plasmid vector is cut at the unique Pst-1 site in the ampicillin resistance gene, and the protruding 3' ends are tailed with dG residues in a similar fashion. It is then possible to anneal the cDNA and plasmid to reform a circular recombinant DNA molecule. In theory, the ends of the tailed plasmid cannot rehybridize. Transformation of a suitable host *E. coli* strain with the recombinant plasmid will confer tetracycline resistance on the host, but will no longer confer ampicillin resistance, due to insertion of the cDNA into this gene. The host cell repairs the gaps at each end of the insert, forming a Pst-1 site at each end. This then provides a convenient method for cutting out the cDNA insert after replication and isolating it from the plasmid.

Thousands of individual bacteria can be transformed by this procedure. Selection for the appropriate antibiotic resistant/sensitive cells enables suitable transformants to be identified. Each one carries a recombinant DNA plasmid into which is inserted a different cDNA sequence. Clones can easily be isolated by plating the colonies. The major task is to catalogue the cDNA library and identify clones.

Alternative methods of cDNA cloning

The scheme shown in Figure 2 is only one of a number of possible cloning procedures. For example, in step 1, it is now possible to anneal the mRNA directly to a plasmid construct containing an oligo (dT) primer for first strand cDNA synthesis[16]. After (dC) tailing of the 3' end of the single strand cDNA, a (dG) tailed linker inserted into the plasmid is then used to prime the second strand synthesis.

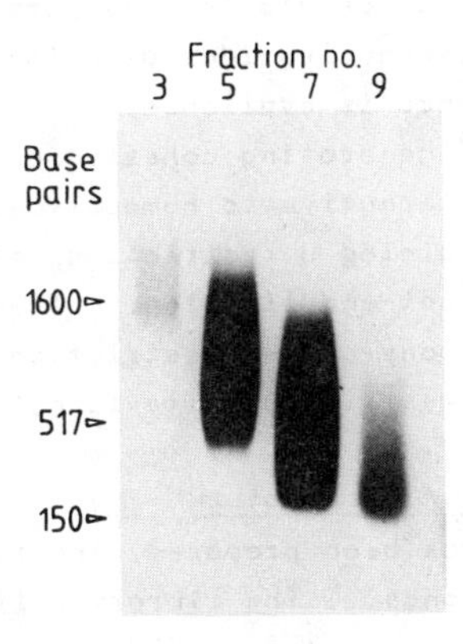

Figure 4 Size fractionation of cDNA

Double stranded cDNA was prepared from a plant poly(A)$^+$ mRNA population as in Figure 3. Following S_1 nuclease digestion the cDNA was size fractionated by passing down a small gel filtration column. The size of molecules in alternate fractions was determined by agarose gel electrophoresis and compared with markers of known molecular weight. This allows selection of larger molecules, which were more likely to be full length transcripts, for subsequent cloning.

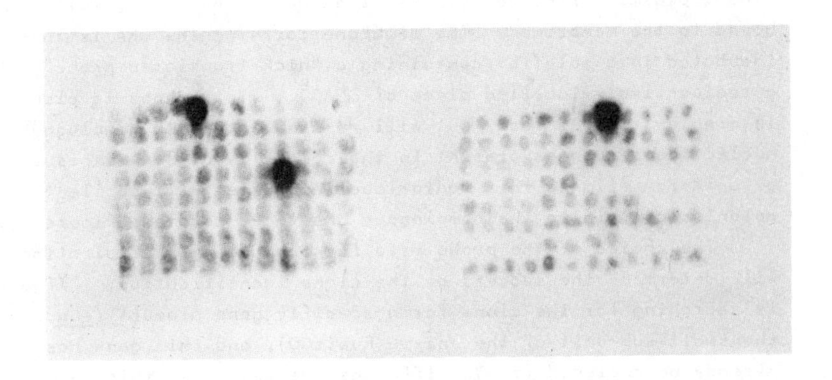

Figure 5 Colony Hybridization

DNA from the cDNA library on gridded filters are probed with ^{32}P-labelled cDNA from one clone. Three clones in this library are shown to contain homologous sequences.

A quite different method of cloning one particular mRNA sequence using specific oligonucleotides as primers is possible if a partial protein sequence is available.

At the later stage of generating cohesive ends on the double stranded cDNA an alternative to homopolymer tailing is to add a DNA oligomer containing a restriction enzyme site to the ends of the cDNA by blunt-end ligation. This then allows insertion of the cDNA into any chosen restriction site in a vector, provided that the restriction site is not also present within the cDNA sequence.

cDNA identification by colony hybridization

Once a cDNA library has been prepared, the first task is to identify the individual clones. The library will probably consist of several hundred or even thousand clones. The usual means of identification is by colony hybridization techniques[17]. The clones are individually spotted onto filter membranes made of nitrocellulose or a similar material, which are laid upon nutrient agar. The clones are grown up to yield small colonies which are then lysed *in situ*, and the released DNA (consisting of the bacterial chromosome plus several copies of each recombinant plasmid) rendered single-stranded, by heat or alkali, and bound to the membrane. The membrane carrying the DNA is then incubated in a solution containing a "nick-translated probe", a radioactively-labelled piece of DNA[18]. This probe is also in single-stranded form, but will re-associate with homologous nucleotide sequences present in the colony lysates. Exposure of an X-ray film to this radio-labelled filter will enable the colonies containing the homologous DNA to be located (Figure 5).

The choice of the probe used for the colony hybridization will determine the success of the clone identification. If one is searching for the clone for a specific gene product (*e.g.* the small sub-unit of the enzyme RubisCO), and this gene has already been cloned from a different species, then this characterised clone can be used as a "heterologous probe". If the two species are related, or the gene is likely to be conserved, the probe will hybridize adequately to the related DNA in the clone library, and the cDNA clone can be "fished out".

An alternative approach depends on the gene product being a protein which has been characterised to the extent of a partial amino acid sequence. The nucleotide sequence coding for this stretch of peptide is deduced and a synthetic oligonucleotide is prepared either chemically or enzymically[19]. In fact several oligonucleotides may be necessary to accommodate the redundancy of the genetic code. This may be minimised by judicious choice of the portion of the polypeptide sequence so that it contains amino acids with unique codons (Met,Trp) or minimal ambiguity (His,Phe,Tyr). A "12mer" (a 12 nucleotide oligomer) may be sufficiently specific to use as a radiolabelled probe under optimal hybridization.

If the required gene product has not been characterised at the molecular level, the process of clone identification is more difficult. Many of the clones in the library may be eliminated by the process of "differential hybridization". The library is screened in duplicate with probes consisting of (i) the cDNA used to prepare the library, and (ii) cDNA prepared from tissues not expressing the genes of interest. Those clones which hybridize equally to both probes can be discarded, and those which hybridize only (or more strongly) to probe (i) are studied more closely (Figure 6). The probes used could be from two different tissues, or from the same tissue, but in different states of differentiation or after different stimuli had been applied (e.g. ripe/unripe fruit, plus/minus hormone, infected/healthy, heat shocked/unshocked). To narrow the field even further one probe could consist of cDNA prepared from a mutant tissue lacking expression of only one or a few genes, or from a tissue where expression has been chemically blocked.

The library may still contain a large number of clones, but many of these will be duplicates, derived from individual species of mRNA. These groups can be categorised by using the cDNA insert from each of the clones as a nick-translated probe (Figure 5). This technique will however also group together cDNAs derived from similar, but distinct messages. If the tissue was predominantly expressing a single gene or a relatively small number (e.g. globin, ovalbumin, seed storage proteins) - a process which may have been enhanced by the use of elicitors prior to RNA extraction - a tentative identity may be assigned to any large group of clones. Also the size of the cDNA cloned in each plasmid will give some indication of the molecular weight of its protein product (if the reverse transcription process was efficient and yielded "full-length"

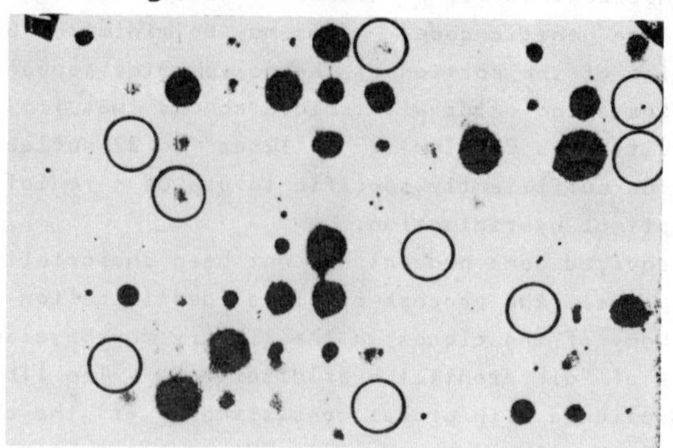

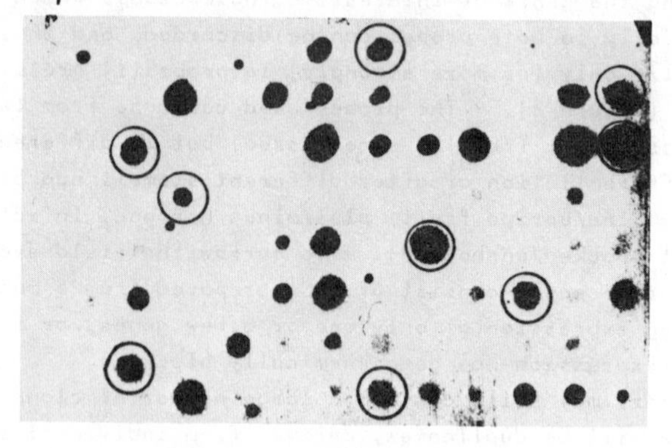

Figure 6 Differential colony hybridization

In this example, cDNA clones prepared from ripe tomato poly(A)$^+$ mRNA were grown in identical positions on two filters ("replica plating") and probed with either ^{32}P-labelled cDNA prepared from mature green fruit poly(A)$^+$ mRNA or ripe fruit poly(A)$^+$ mRNA. Circled colonies indicate cDNA clones which hybridize more strongly to ripe fruit cDNA than to mature green fruit cDNA and presumably contain mRNA sequences which increase during ripening (see Figure 1).

cDNAs) and this can be compared to the known size of any proteins produced *in vivo* by the tissue.

Characterisation of cDNA clones by expression

Two distinct methods have been used to identify the proteins encoded by individual cDNAs, using cell-free protein synthesis (Figure 7). In both methods the principle is the same in that the hybridization of cloned cDNA to its complementary mRNA can be detected by translation *in vitro*. Hybrid arrested translation (HART) relies on the ability of the cDNA-mRNA hybrid to specifically inhibit translation of only that particular mRNA in the whole population. Thus, after hybridization, one band would be missing from the *in vitro* translation profiles of the mRNA sample, which would reappear after dissociating the hybrids by heating before translation. Obviously, this method only works if the translation product can be resolved by gel electrophoresis (Figure 7).

The alternative method involves binding the clones cDNA to a solid support (*e.g.* DBM paper or nitrocellulose membrane) and hybridizing total mRNA to this. After vigorous washing, the hybridized mRNA can be eluted by disrupting the hydrogen bonding, and translated. In this case, one band only, corresponding to the specific mRNA translation product, would be observed (Figure 7). The identity of a hybrid-selected mRNA can be established by immunoprecipitation of its translation product, or by partial proteolysis, followed by "fingerprinting" of the polypeptide fragments.

In some cases direct identification of the cDNAs can be achieved by cloning them into an expression vector. This is a vector carrying all the signals necessary for expression of a gene in the bacterial host just upstream of the point where the cDNA is inserted (Figure 8). They must contain a promoter region, such as the P_L promoter from phage λ, the *trp* or the *lac* promoter from *E. coli*, or a synthetic hybrid of *trp-lac* (*tac*), and also a ribosome-binding site for efficient translation of the message[20]. The cDNA is then transcribed in the host, and provided the mRNA is stable, it is translated to yield a protein which can then be detected either by enzyme assay or immunoassay. Often the vector contains the N-terminal sequence of a bacterial gene (such as the *lacZ* gene which codes for β galactosidase) and the cDNA-encoded protein is fused to this. In this case the

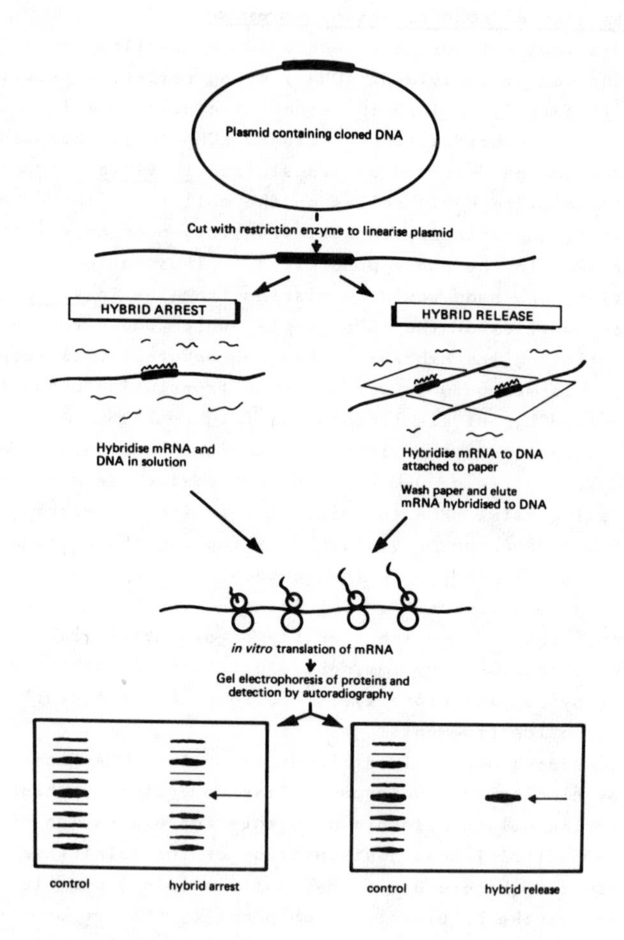

Figure 7 Hybrid selected translations

A cDNA clone is hybridized to its homologous message in a mixture of mRNAs. This will prevent translation of that particular mRNA (Hybrid Arrest). Alternatively the mRNA can be eluted and translated to give a single identifiable product. (Figure reproduced from reference 25 with permission).

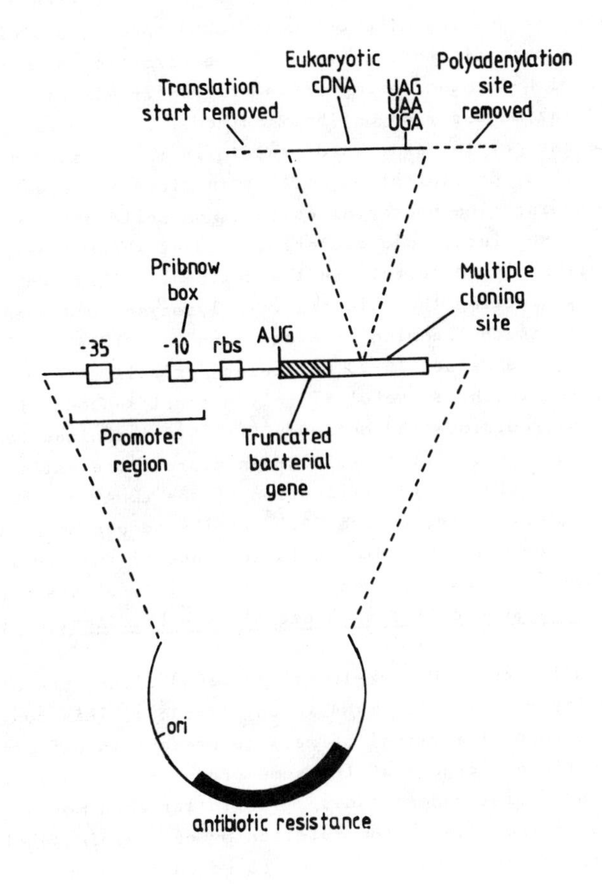

Figure 8 A schematic vector for the expression of eukaryotic genes in bacterial cells

The vector, based on a high copy number bacterial plasmid carries a series of cloning sites capable of accepting foreign DNA cut with a variety of restriction enzymes. 5' to this site are the bacterial promoter sequences, a binding site for 60S ribosomes, and the N-terminal portion of a bacterial gene. Translation of this gene then reads through and the foreign DNA is expressed.

vector is normally available in three forms, differing by the presence of an additional nucleotide or two at the fusion point to ensure that in one case the truncated cDNA-encoded protein will be in the correct reading frame. The formation of proteins with additional N-terminal amino acids frequently allows detection by enzyme assay or antibody methods.

The presence of the expressed protein is most easily detected by antibody precipitation, either in micro-titre wells or directly against the bacterial colonies on solid media. If the protein is not fused to a bacterial product which is exported, the cells must be lysed to release the antigen. This can easily be achieved by treating the colonies with lysozyme, and then sarkosyl, prior to challenging with antisera[21]. This system is sensitive enough to detect 10-20 molecules per cell. A more sensitive system which is useful if only a small sequence is expressed, uses radiolabelled antibody[22]. This technique can also be used *in situ* on solid media or in micro-titre wells.

Finally, unequivocal identification of the cDNAs can be achieved by nucleotide sequencing[23, 24]. The sequence obtained can then be compared to the amino acid sequence of the required protein product (if known) to confirm the identity of the clone.

Dot-blotting: measurement of mRNA amount by hybridization to a cDNA clone

Isotopically-labelled nick-translated cDNA clones can be used to quantify a particular mRNA in any tissue. This can give a comparison of the amount of message present in different species, in different organs of the same organism and in various mutants of the species under study. The latter case may provide some evidence of the site of the mutation by determining whether the message is absent, reduced, or increased in the mutant organism. Another simple use of the dot-hybridization technique is to take RNA samples at various times during a developmental process (ripening, physical stress, heat, osmotic, light/dark, wounding, parasitic invasion) and to challenge this with the cDNA clones to detect any progressive change in the level of the specific mRNA and then correlate this with the biochemical changes observed.

The method used is as follows:- RNA is prepared from the tissue under study and the mRNA fraction may be purified by passage through an oligo-dT cellulose column. The RNA is denatured (with glyoxal or heat) and then spotted on to a nitrocellulose or nylon membrane filter, usually with the aid of a manifold and a vacuum. A series of standard dilutions of RNA or DNA may also be applied to quantify the results. After baking the membrane carrying the RNA, it is incubated in a solution containing an oligonucleotide mixture (e.g. denatured salmon sperm DNA, yeast tRNA) which binds non-specifically to the filter and the RNA. The nick-translated cDNA clone is then denatured and introduced to the mixture, and it will hybridize to its complementary mRNA. Some hours later the membrane is washed to remove non-specifically bound DNA, dried, and autoradiographed. The resulting X-ray film can be scanned with a densitometer to determine the amount of radioactive probe bound to each "dot". Alternatively the dots can be located on the original membrane, cut out, and counted in a scintillation counter (Figure 9).

Measuring the molecular weight of a mRNA by Northern blotting

A technique similar to dot hybridization can yield further data on the expression of the genes under study and the mRNA species present. This is Northern hybridization. In this method, the RNA from the tissue under study (either total RNA or $PolyA^{+}$-containing mRNA) is denatured and then separated electrophoretically in an agarose gel. The RNA is then transferred from the gel to a membrane filter by placing one upon the other and allowing the transfer to proceed by capillary action or by electrophoresis. The filter is then baked, prehybridized, hybridized, and autoradiographed as before.

Because the RNA has been separated by electrophoresis prior to "blotting" this technique can give some data on the size of the mRNA complementary to the cDNA clone. To calculate this, standard DNA fragments can be run on the gel as markers, or the sizes of the ribosomal RNAs used in the calculation. This will give some reciprocal information on the efficiency of the construction of the cDNA library, by showing whether the cDNA insert in the clone used is of comparable size to the homologous message. It will also reveal the quality of the RNA preparation by highlighting any breakdown products in the gel. If this latter effect is not great it may be possible to detect one or more mRNA species which are homologous to the probe used. This might indicate the expression of several very similar genes or differential processing of the RNA.

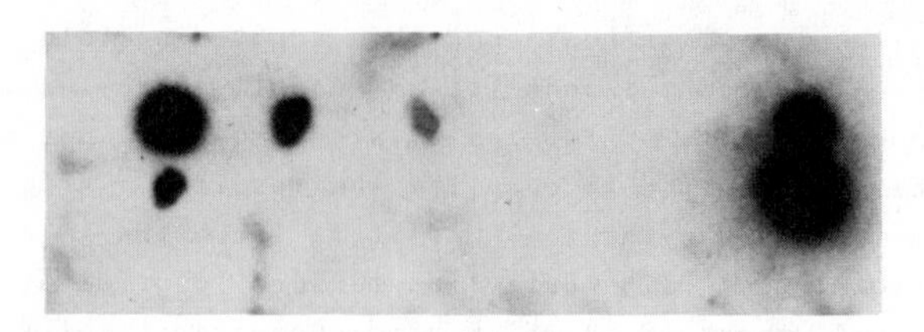

Figure 9 RNA dot hybridizations

A series of diluted RNA dots (1, 0.1, 0.01, 0.001 and 0.0001 μg) from two different tissues are probed with a ^{32}P-labelled cDNA clone. The RNA from the first (upper) tissue contains approximately 30x more homologous mRNA than that from the other. Standard amounts of DNA are included on the right hand side to quantify the result.

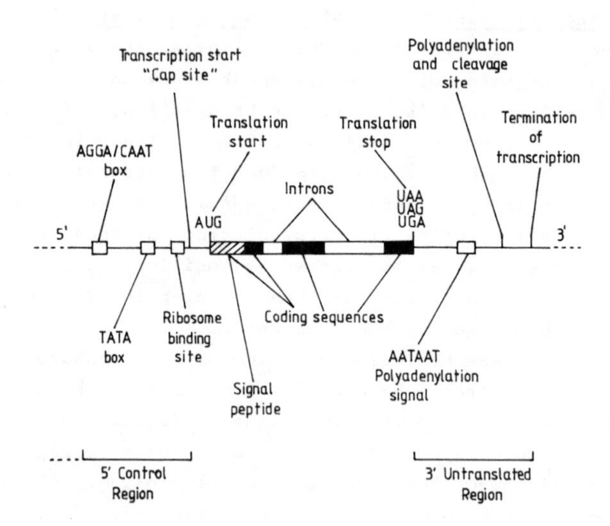

Figure 10 A eukaryotic gene

The protein-coding portions of the gene (possibly including that for a signal peptide) are interrupted by non-translated introns. Signals in this region are marked as they would appear in the corresponding mRNA (i.e. U substitutes for T). 5' to the structural gene is the control region or promoter and ribosome binding site. Downstream (3') are the signals for polyadenylation of the mRNA, the site where this occurs, and the termination point for transcription.

If the dot blot technique showed the presence of mRNAs in different species, mutants, or tissues, Northern analysis may be able to distinguish differences, for example, because the mRNAs may be of different size. This would indicate, in the case of a different species, the expression of a similar but distinct gene coding for the mRNA. In different tissues, different sizes of mRNA could be due to processing differences, or the expression of different genes. The same is true in the case of mutant tissue, but gene rearrangement leading to the production of "nonsense transcripts" may also occur. These show up in Northern hybridization experiments but are not translated correctly by protein synthesis *in vitro*. If the messages are of similar size in both mutant and normal tissue, this would indicate either that the mutation does not involve this gene, or that the effect of the mutation occurs later in the translation/ protein maturation stages.

Studies on RNA transcription and processing Changes in the amount of specific mRNAs occur in many biological situations. This could result either from enhanced mRNA transcription or a decreased rate of degradation. It is, therefore, important to establish whether or not the appearance of a mRNA is due to an alteration in transcription as a preliminary to studying the regulation of a particular process. Transcriptional control can be tested by studying RNA synthesis by isolated nuclei. These are isolated from cells or tissues both before and after they initiate the process thought to involve transcription of a mRNA for which a cDNA clone is available. The nuclei are incubated under conditions where RNA synthesis can proceed *in vitro*. At the end of the incubation the RNA is extracted and fractionated by electrophoresis, transferred to a membrane by Northern blotting, and the membrane is probed by hybridization with a radioactive cDNA. RNA samples from nuclei actively transcribing the gene corresponding to the cDNA probe hybridize strongly, whereas RNA from nuclei inactive in transcription of the same gene shows little or no hybridization.

Studies with nuclear RNA, synthesised *in vivo* and *in vitro*, can also give information about stages in the processing of mRNA molecules. Since the primary transcription product of a gene generally undergoes "processing" (*i.e.* the removal of terminal and internal sequences, the ligation of the conserved regions, and the addition of specific features such as a 5' cap and a 3' poly(A) tail, analysis of Northern blots of nuclear RNA can be used to provide information about the molecular weight and amount of various intermediates in the processing pathway.

Detection of genomic sequences by Southern blotting

The cDNA clones can also be used as probes against genomic DNA from the species of interest. The genomic DNA (total, nuclear or organellar as required) is digested by a certain restriction enzyme (or a combination of several) and then separated by electrophoresis in an agarose gel. Following denaturation in alkali the DNA is transferred to a membrane in a similar fashion to the "Northern blotting" technique and hybridized to the isotopically labelled cDNA in the same fashion.

The resultant X-ray film will locate the DNA band or bands present in the digest which are homologous to the cDNA clone. In this way it is possible to determine whether several different genes are related to the cDNA or whether it has been copied from mRNA expressed from a unique or repetitive sequence. If the latter is the case it can also be determined how many copies of this gene are present in the genome by incorporating a "copy number standard" piece of DNA on the gel. By comparison of the different band sizes obtained using different restriction enzymes an approximate physical map of the gene can be drawn up. Comparison of this with the known restriction map of the cDNA may reveal differences which can be accounted for by the presence of such features as introns.

Needless to say the features described above may differ between different species or mutants, but should be identical within the same organism. However, differences may be noticed between tissues in varying stages of differentiation which reflect differences in "expressibility" of the DNA. Using restriction enzymes which are incapable of digesting methylated DNA (e.g. Sst II, Cla I, Hpa II, Msp I, Pvu II etc.) it may be observed that the DNA is more refractory to digestion, and by inference more highly methylated in a cell in a certain state of differentiation (or mutation). A high degree of methylation is often associated with a reduction of expression of the gene in question. Similarly, if chromatin is prepared from isolated nuclei and then subjected to limited digestion by nucleases prior to electrophoresis, Southern blotting, and hybridization, regions of the genome which are actively expressing will be accessible to digestion and will show a series of bands, whereas "silent" regions will be inaccessible and will appear as high molecular weight fragments.

Identification of genomic clones

One of the more obvious applications of cDNA clones is to identify the homologous genomic clones from a genomic DNA library. The latter is usually constructed in a bacteriophage λ or cosmid vector which can contain approximately 20-45 kilo base pairs of insert DNA. The relevant clones are identified by colony hybridization to the cDNA probe.

Once identified the genomic clone can be analysed by restriction mapping and sequencing. Comparison with similar data from the cDNA clone will reveal the exact position of features such as introns and polyadenylation signals, and may locate others such as the TATA, CAAT and AGGA boxes. Hybridization of the genomic clone to total RNA samples followed by S_1 nuclease digestion will enable the location of transcriptional start sites (Figure 10, p.186).

A further important reason for studying genomic clones is to provide information about control regions thought to be involved in regulating expression. Nearly all eukaryotic cells contain a complete set of genes required for regeneration of a new individual. These are not normally all expressed and specific cell types differentiate by a process involving the regulated expression of specific genes. Control regions exist in the DNA surrounding the actual sequence coding for a protein and these are thought to interact with regulatory factors that stimulate the expression of particular genes (Figure 10). This area of research is currently of intense interest since it may help to unravel a range of problems related to development, disease and food production. Since these control regions are often not transcribed as part of the mRNA, they are not present in the cDNA. The only way of studying their structure and function, therefore, is to seek out the parent genomic sequences and investigate their structure and function.

[1] J.M. Chirgwin, A.E. Przybyla, R.J. MacDonald and W.J. Rutter, Biochemistry, 1979, 18, 5294

[2]V. Glisin, R. Crkvenjakov and C. Byus, Biochemistry, 1974, 13, 2633

[3]H. Aviv and P. Leder, Proc. Natl. Acad. Sci. U.S.A., 1972, 69, 1408

[4]M. Adesnik, M. Salditt, W. Thomas and J.E. Darnell, J. Mol. Biol., 1972, 71, 21

[5]D.J. Shapiro, J.M. Taylor, G.S. McKnight, R. Palacios, C. Gonzalez, M.Z. Kiely, R.T. Schimke, J.Biol. Chem., 1974, 249, 3665

[6]J.M. Taylor, Ann. Rev. Biochem., 1979, 48, 681

[7]H.R.B. Pelham and R.J. Jackson, Eur. J. Biochem., 1976, 67, 247

[8]P.H. O'Farrell, J. Biol. Chem., 1975, 250, 4007

[9]D. Baltimore, Nature, 1970, 226, 1209

[10]H. Temin and S. Mizutani, Nature, 1970, 226, 1211

[11]E.F. Retzel, M.S. Collet, A.J. Faras, Biochemistry, 1980, 19, 513

[12]S.L. Berger, D.M. Wallace, R.S. Puskas and W.H. Eschenfeldt, Biochemistry, 1983, 22, 2365

[13]A. Efstratiadis, F.C. Kafatos, A.M. Maxam and T. Maniatis, Cell, 1976, 7, 279

[14]R. Higuchi, V.G. Paddock, R. Wall and W. Salser, Proc. Natl. Acad. Sci. U.S.A., 1976, 73, 3146

[15]F. Bolivar, R.L. Rodriguez, P.J. Greene, M.C. Betlach, H.L. Heyneker, H.W. Boyer, J.H. Crossa and S. Falkow, Gene, 1977, 2, 95

[16]H.Okayama and P. Berg, Molec. Cell. Biol., 1982, 2, 161

[17]M. Grunstein and D.S. Hogness, Proc. Natl. Acad. Sci. U.S.A., 1975, 72, 3961

[18]P.W.J. Rigby, M. Dieckmann, C. Rhodes and P. Berg, J. Mol. Biol., 1977, 113, 237

[19]J.W. Szostak, J.I. Stiles, B-K. Tye, P. Chiu, F. Sherman and R. Wu, Meth. Enzymol., 1980, 68, 419

[20]J. Shine and L. Dalgarno, Nature, 1975, 254, 34

[21]A. Skalka and L. Shapiro, Gene, 1976, 1, 65

[22]H.A. Ehrlich, S.N. Cohen and H.O. McDevitt, Cell, 1978, 13, 681

[23]A.M. Maxam and W. Gilbert, Proc. Natl. Acad. Sci. U.S.A., 1977, 73, 560

[24]F. Sanger, S. Nicklen and A.R. Coulson, Proc. Natl. Acad. Sci. U.S.A., 1977, 74, 5463

[25]D. Grierson and S. Covey, "Plant Molecular Biology", Blackie, Glasgow, 1984.

10
Enzyme Technology: towards Usable Catalysts

By M. D. Trevan

DIVISION OF BIOLOGICAL AND ENVIRONMENTAL SCIENCES, THE HATFIELD POLYTECHNIC, P.O. BOX 109, COLLEGE LANE, HATFIELD, HERTS. AL10 9AB, U.K.

1. ENZYMES AS CATALYSTS

The application of enzymes is on the one hand a major concern of modern biotechnology and on the other is a pursuit as old as man's civilisation. Enzymes have been used for centuries in processes such as cheese making or the leather industries and even today it is these traditional uses of enzymes which predominate. The purpose of this text is briefly to consider the scope of applied enzymology and then to concentrate on some of the presently or potentially available techniques which might, in future, extend the range of applications.

It was estimated that in 1981 the world market for commercially produced enzymes was around 65,000 tonnes valued at $\$400 \times 10^6$. These are figures for enzyme sales alone. At a conservative estimate this probably represents some $\$24 \times 10^9$ worth of product produced by or incorporating enzymes worldwide. Thus a fairly substantial amount of economic enterprise depends upon the humble enzyme.

Why is there so much interest in the application of enzymes? The answer lies in the unique catalytic abilities of enzymes: their enormous catalytic power and their high specificity for both the reaction catalysed and the substrate converted. The catalytic power of an enzyme can be illustrated by the catalase (E.C.1.11.1.6) which catalyses the reduction of hydrogen peroxide to water. The active ingredient of the enzyme is a single atom of iron located at the active site. In this catalase is not unique; iron filings will also catalyse the breakdown of hydrogen peroxide, but to achieve a rate of breakdown obtainable from 1 mg of iron in catalase takes several tonnes of iron filings. Enzymes typically have turnover numbers (that is the number of substrate molecules processed by one enzyme molecule per second) in the region of 10^4, with some highly efficient enzymes (e.g. acetylcholinesterase E.C.3.1.1.7) having turnover numbers of 10^8. These super enzymes are so near to perfection that the rate of reaction is effectively controlled by the rate of collision between substrate and enzyme molecules. Furthermore their catalytic power is evident at ambient temperatures and pressures. Enzymes are highly specific for their substrate molecules usually exhibiting the ability to distinguish between stereoisomers of the substrate, to a degree which often eludes the organic chemist.

Some 2,500 different enzymes have been isolated and described to date and it is estimated that this probably reflects only 10% of the enzymes existing in nature.

It is of little surprise, therefore, that enzymes are widely used as analytical reagents, for therapeutic purposes (e.g. the use of urokinase (E.C.3.4.31.21) to prevent the occurrence of pulmonary embolisms), as manipulative enzymes (e.g. the preparation of protoplasts or the manipulation of DNA) and as industrial catalysts (e.g. the isomerisation of glucose syrups to produce high fructose syrups). Some of these applications are listed in Table 1. Even so, probably fewer than 200 enzymes are used in such applications.

The purpose of this treatise is to pose two questions. Why do relatively so few enzymes find practical application? What can be done to extend the range and scope of enzyme applications?

2. LIMITATIONS TO ENZYME APPLICATION

There are a number of factors which have limited the range of enzyme applications, particularly in their use as industrial catalysts; availability, cost, and, perhaps most significantly, stability.

2.1 Availability

It is perhaps surprising that enzyme application should be limited by availability, given the vast range of enzymes in nature. The problem is both practical and legislative. When considering the manufacture of a particular enzyme, an enzyme producer must, in order to reduce costs by bulk production,

TABLE 1. Some Applications of Enzymes

Enzyme	Application	Form of Enzyme
Analytical		
Glucoseoxidase	Detection of D-glucose in blood samples	Soluble. Immobilised in coil
Luciferase	Bacterial contamination of food/water	Soluble
	Viability of bull semen	Soluble
Peroxidase	Quantification of hormones or antibodies	Linked to immunoglobin
Therapeutic		
Asparaginase	Treatment of some leukaemias	Immobilized in liposomes
Urokinase	Preventing of pulmonary embolism	Soluble
Manipulative		
Lysozyme	Bacterial cell lysis	Soluble
Nucleases	Genetic manipulation	Soluble
Industrial		
Amylases	Conversion of starch to glucose or dextrins in food industries, e.g. baked goods to increase sugar content for fermentation; brewing removal of starch turbitity; confectionery recovery of sugar from waste	Soluble
Catalase	Conversion of latex to foam rubber	Soluble
Lactase (β-galactosidase)	Prevention of crystalisation of lactose in ice cream	Soluble
	Conversion of lactose in cheese whey	Immobilized
Nitrilase	Manufacture of acrylamide	Immobilized
Proteases	Laundryaid; Recovery of silver from spent photographic film; Improvement of loaf volume	Soluble

manufacture the enzyme in a quality which is acceptable to all of his customers. In practice this means that the enzyme will often be produced as a single grade to satisfy the most stringent legislative demands, as it will then perform satisfactorily where lower quality will suffice. The most stringent regulations are, of course, for food grade enzymes. Where the enzyme is derived from a microbial source, this immediately limits production to a food compatible microorganism.

Thus most industrial enzymes are produced from only 11 fungi, 4 yeasts and 8 bacteria. Any new enzyme required is first sought from these organisms. This approach has an additional advantage in that the producer has to hand the knowledge and technology to grow and extract enzymes from these microbes.

The disadvantage of this philosophy is, of course, that the isolated enzyme may not show ideal characteristics for its intended purpose, for example substrate specificity or stability may not be as high as that for a similar enzyme from an untried microorganism.

So why use microbes to produce enzymes, why not use plant or animal materials? The answer lies in part in the ease with which fluctuations in demand for a particular enzyme can be catered for by altering fermentation capacity, partly in the wide range of reactions exhibited by most microbial cells and in part the convenience and certainty of not having to rely on an external supply of raw material.

To the enzyme user all this may restrict the choice of enzyme, but will assure that the enzyme will be reliably available in bulk and of a known quality. What redress does the enzyme user have if a suitable enzyme is not available? He may attempt to alter the characteristics of an available enzyme; look for new reactions catalysed by existing enzymes; select a new source and do it himself. All three approaches have been researched in recent years with some interesting results.

Selecting a new source of organism and alteration of the characteristics of available enzymes are discussed below in section 4 and elsewhere in this volume.

The search for new reactions from old enzymes is a relatively new field of endeavour but already there are some surprises. For example, glucose oxidase (E.C.1.1.3.4), normally associated with the catalysis of glucose and oxygen to gluconolactone and hydrogen peroxide, has been employed to covert benzoquinone to hydroquinone in the presence of glucose. Carboxypeptidase A (E.C.3.4.17.1) has two activities, esterase and peptidase, and normally contains a zinc atom. If this is replaced by manganese or cadmium the peptidase activity is virtually lost whilst the esterase activity is enhanced. α-thrombin (E.C.3.4.21.5) also has two enzyme activites, amidase and esterase, the balance between which can be altered by adjusting the level of dimethyl suphoxide. Proteases in general may be used to synthesise esters or peptides, reversing the normal hydrolytic reaction by performing the reaction in organic solvents. Thus α-chymotrypsin (E.C.3.4.21.1) in chloroform will synthesise

N-acetyl L-tryptophan ethyl ester, with 100% yield, from ethanol and N-acetyl L-tryptophan; use of carbon tetrachloride as solvent lowers the yield to less than 60%. This approach to extending the availability of enzyme catalysts is reviewed more extensively by Neidleman (1984)[1].

2.2 Cost

In many applications of enzymes, particularly industrial, cost of the enzyme may be of paramount importance. Even the cheapest of bulk enzymes, e.g. alkaline proteases for washing powder formulation, may contribute a significant fraction of the final product value, in this case about 5-10%. The situation is exacerbated by the traditional use of enzymes as soluble reagents in industrial and other processes, resulting in loss of the enzyme at the end of the process. For an inexpensive enzyme this may not matter; however for an expensive enzyme single use may be prohibitively costly.

Clearly a reusable enzyme would be advantageous; the problem to overcome is enzyme solubility, the solution is to render the enzyme insoluble. This may be easily done by fixing or immobilizing the enzyme onto an inert, insoluble polymer molecule. The immobilized enzyme may then easily be retained within the 'reactor' (or analytical device). Much work and effort has gone into studies of the application of immobilized enzymes (and cells) over recent years and many rewards are still to come. For a full discussion of enzyme immobilization see Klibanov (1983)[2] or Trevan (1980)[3].

Quite obviously, however, a reusable enzyme will be useless unless it is stable enough to permit reuse.

2.3 Stability

Enzyme stability, or rather lack of it, is one of the key factors hampering wider application of enzymes. When considering stability of enzymes we are effectively concerned with an enzyme's resistance to denaturation by physical, for example heat or pH, or chemical factors, for example urea or organic solvents. In general, enzymes which are stable at high temperatures may often be stable in organic solvents or at extremes of pH, and it is for this reason that most of our discussion will be centred on heat and to a lesser extent organic solvent stability.

The need for heat stability is self evident: the more stable the enzyme the longer it will last in a reactor. In addition thermostable industrial enzymes allow the possibility of carrying out processes at elevated temperatures, reducing the risk of microbial contamination and markedly elevating reaction rates. This latter point in particular aids process economics, as it allows either a smaller enzyme reactor or a greater productivity. However, there are other advantages in having a thermostable enzyme. Shelf life of the enzyme will be enhanced, enabling batch production (of enzyme) to be performed on a larger, more cost effective scale, as well as promoting user convenience. Thermostable enzymes will be less prone to denaturation by the conditions under which they are produced, which may in turn lead to increased yields of enzyme or permit the use of simpler and cheaper production methods.

The rationale for organic solvent stable enzymes has been hinted at above (reverse hydrolysis) but there is another, wider reason. Many interesting and valuable substances which could be produced from enzyme catalysed processes are too insufficiently soluble in water to make process design feasible. The ability, therefore, to run the reaction in the presence of organic solvents would be valuable in the processing of molecules such as steroids or triglycerides.

3. ENZYME DEACTIVATION

One of the problems of studying enzyme stability has been the lack of suitable models of enzyme deactivation. Generally deactivation has been considered to be a first order process, which although successfully applied to a number of enzymes could not explain the deactivation of many others. Equally, simple deactivation models also gave rise to a number of dubious claims for enhancement of enzyme stability by this or that procedure. This lack of understanding has probably hindered the application of rational approaches to enzyme stabilization, as has a lack of knowledge of the molecular forces responsible for enzyme stability. Thus, to this time, most attempts at enzyme stabilization have been empirical in nature and often with dubious results.

Happily, the situation is changing with a number of recent articles on both these subjects [4,5]. Henley and Sadana (1985)[5] describe a two step model for deactivation, involving the irreversible interconversion of three conformations of the enzyme molecule.

$$E \xrightarrow{k_1} \overset{\alpha_1}{E_1} \xrightarrow{k_2} \overset{\alpha_2}{E_2}$$

where k_1 and k_2 are first order deactivation rate constants, E, E_1 and E_2 are specific activities of the three conformational states of the enzyme, and $\alpha_1 = E_1/E$ and $\alpha_2 = E_2/E$.

This model yields an equation relating the total enzyme activity to the values of α_1, α_2, k_1 and k_2 and time. They then went on to produce model curves (by altering these values) which fitted known experimental data, and found that the experimental data could be divided into 14 cases, broadly classified into two categories, the first where the activity (a) at any time never exceeds the initial activity (a_0) and the second, intriguingly, where (a) can exceed (a_0), that is the enzyme, at least transiently, becomes more active.

The lesson to be drawn from these and similar studies is that enzymes which have been modified may indeed show enhanced stability over the native enzyme by virtue of the fact that the modification may have altered the balance between co-existing conformations of the enzyme. Consider the case where $k_1 > k_2$ and $\alpha_1 < 1$ but $> \alpha_2$. The native enzyme will exhibit rapid initial deactivation followed by a slower deactivation rate (fig. 1). If modification of the enzyme substantially denatures E but not E_1 then the modified enzyme will apparently display greater stability. This mechanism is seen in the deactivation of trypsin (E.C. 3.4.21.4) which exists in two forms α and β differing only in the extent of cleavage of the peptide chain. The rate constant of deactivation of the α form is 100 times that of the β form[6].

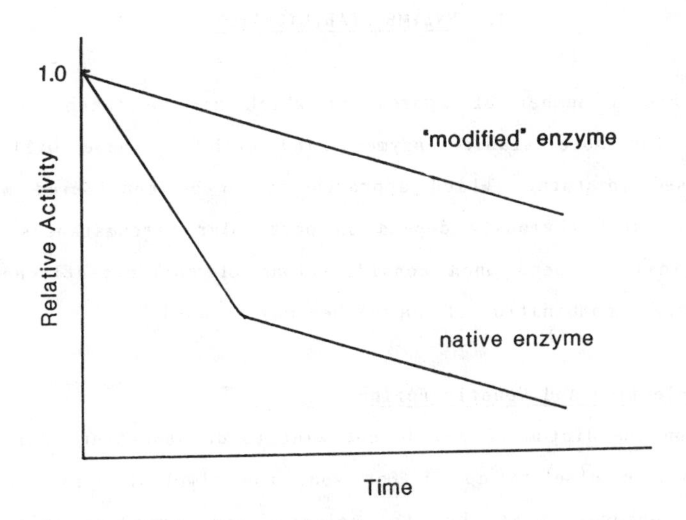

Fig.1 ***Artefactual stabilisation by differential denaturation of coexisting enzyme conformations***

In addition to the deactivation of enzymes brought about by an unfolding of their molecular structure, there are a number of other reasons why enzymes may become deactivated during use. For example, organic solvents, in addition to their ability to denature enzymes, can cause the aggregation or precipitation of an enzyme with consequent loss of activity; microbes may act to 'digest' the enzyme; the presence of contaminating proteases may hydrolyse the enzyme; if the enzyme is itself a protease it may suffer autolysis; the enzyme may be poisoned by the presence of, for example, metal ions or even oxygen. Such effects may be simply countered by making the enzyme inaccessible to the inactivating agent by immobilizing the enzyme (q.v.).

4. ENZYME STABILIZATION

There are a number of approaches which may be taken in the search for more stable enzymes, and each of these will be discussed in turn. Which approach is to be considered most suitable will obviously depend on particular circumstances and will finally depend upon considerations of cost/effectiveness. Obviously a combination of approaches may be used.

4.1 Selection and Genetic Engineering

Based on the dictum if you do not want to do something yourself get someone else to do it for you, the simplest approach to stable enzymes must be to select them from a suitable microorganism. In this respect the increasing body of knowledge of thermophilic organisms is proving invaluable. Thermophiles, by virtue of their ability to survive at high temperatures, invariably seem to exhibit enzymes with a high thermal stability. At this time the only major area in which this approach has been successfully applied is the production of thermostable proteases for inclusion in enzyme based detergents. One of the limitations of this approach is, unlike mesophilic organisms, the present inability to persuade thermophilic organisms to overproduce a particular enzyme. This problem may, theoretically, be overcome by transferring the gene for a thermostable enzyme from a thermophile into an appropriate mesophile. However, such studies are still largely undeveloped and few reports exist in the literature.

4.2 Protein Engineering

The subject of protein engineering is discussed elsewhere in this text (see Chapter by J.M. Walker) and it will suffice here to point out that such an approach may, in the longer term, generate a new breed of highly stable enzymes.

4.3 Reaction Environment

There are many examples of instances where the presence of substrate molecules or specific metal ions (*e.g.* Ca^{2+} or Zn^{2+}) has a marked effect on an enzyme's stability. For example, α-amylase (E.C.3.2.1.1) from *Bacillus licheniformis* is stabilized by the presence of 4 p.p.m. Ca^{2+} ions to such an extent that it can retain 100% of its activity after 6 hours incubation at 70°C. In the absence of Ca^{2+} ions it is totally denatured after 4 hours at this temperature. This approach cannot be applied to all α-amylases; that from *Bacillus amyloliquifaciens* shows no increased stability in the presence of Ca^{2+} ions. However, most amylases are stabilized by the presence of their substrate starch. Such effects may be explained by the contribution that these additional bonds make to the stability of the protein's structure or indeed to its native conformation. For example, removal of the haem group from myoglobin causes a decrease in total α-helix content of the protein from 75% to 60%.

Addition of high molecular weight hydrophilic polymers (e.g. dextrans or polyethyleneglycols) to an enzyme solution can stabilize the enzyme[7] either by increasing the viscosity of the solution or by lowering the water activity around the enzyme removing its hydration shell, and thus restricting the enzyme's ability to alter its conformation. Presumably, by removing the ability of the enzyme's hydrophilic groups to interact with water they have no option but to interact with each other, thus preventing an unfolding of the molecule. This form of stabilization is illustrated by α-amylase which in the presence of 80% dry solid starch "solution" can be made to operate effectively at a temperature of 110°C.

The water concentration of an enzyme solution can of course be lowered by other means, often with the same stabilizing effect. Zaks and Klibanov (1984)[8] have reported the stabilization of porcine pancreatic lipase (E.C.3.1.1.3) by decreasing the water content of the system. In an aqueous medium the enzyme instantaneously loses its activity upon heating to 100°C. However, when the dry powdered enzyme was placed in a mixture of 2 mol dm^{-3} n-heptanol in tributyrylglycerol (suitable substrates for the lipase) the half-life of the enzyme at 100^oC was more than twelve hours. In this case the water content of the system was only 0.015%. Increasing the water content to 0.8% caused a reduction in half life to 15 minutes.

However these examples also illustrate the problem associated with attempts to restrict conformational changes in enzymes. In both cases the specific activity of the enzyme was reduced, substrate specificity altered and undesirable by-products formed. In the case of α-amylase a variety of trisaccharides were produced. The stabilized lipase, unlike the native lipase, was less able to utilise bulky secondary and tertiary alcohols, whilst its specific activity at 100^oC was no greater than that for the native enzyme in water at 20^oC.

One final form of reaction environment modification which may successfully increase enzyme stability is to increase the protein concentration. This may function in one of two ways. First, if proteolysis or autolysis is the cause of deactivation, then providing a high concentration of an alternative substrate will reduce loss of the enzyme. Second, physical interactions between proteins are often dependent upon the concentration. Thus high concentrations, whilst apparently reducing the tendency of protein molecules to aggregate and precipitate, may well allow specific, stabilizing interactions to occur (q.v. immobilization). Forniani et al. (1969)[9] showed that human glucose-6-phosphate dehydrogenase (E.C.1.1.1.49) retained 90% of its activity at 37^oC for 90 days at a concentration of 1 unit cm^{-3} but lost 80% of its activity under the same conditions at 0.06 units cm^{-3}.

4.4 Rebuilding

On the theoretical basis that thermostability in enzymes is largely a result of hydrophobic interactions in the care of the enzyme [10, 11], Mozhaev and Martinek (1984)[4] have proposed a route to enzyme stabilization by enhancing these interactions. They suggested that in order to gain access to the hydrophobic groups the enzyme is first unfolded (by treatment with urea and a disulphide) and then refolded after chemical modification, in the presence of low molecular weight ligands which would be incorporated into the centre of the enzyme molecule, or refolded in non-native conditions (e.g. at a higher temperature) (Fig. 2). When they tried the last of these approaches on trypsin, refolding it at 50°C, the subsequent active enzyme was 5 times as stable at 80°C as either the native enzyme or the unfolded enzyme refolded at 20°C. Presumably the enhanced enzyme stability is a result of strengthened hydrophobic interactions brought about by the increased temperature during refolding.

4.5 Chemical Modification

Chemical modification of the surface of an enzyme molecule as a means of achieving stabilization is a well practised art. In theory any modification to an enzyme's primary structure may result in conformational effects which will alter stability. The trick is to ensure that these alterations are beneficial. By virtue of the lack of understanding of the physical relationship between structure and stability this, and other, approaches have been largely empirical, many modifications causing either total loss of activity or reduced stability. However, some encouraging results have been reported.

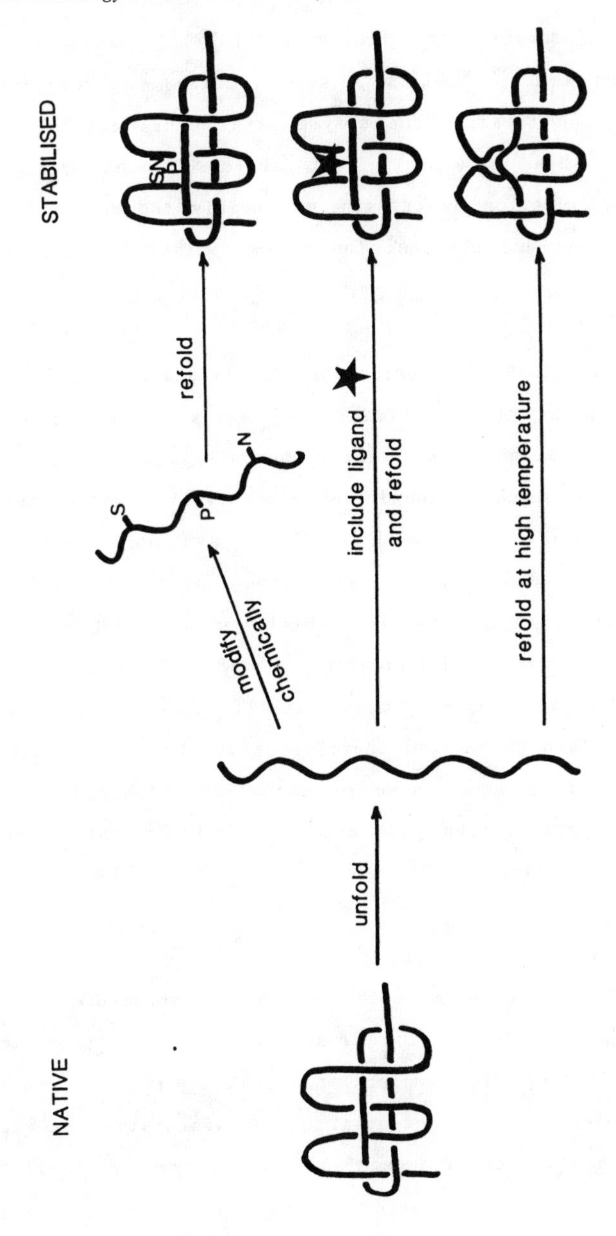

Fig.2 ***Routes to enzyme rebuilding***

For example, Tuengler and Pfleiderer (1977)[12] enhanced the heat and alkali stability of lactate dehydrogenase (E.C.1.1.1.27) by treating the enzyme with methyl acetimidate. Interestingly this treatment also increased the enzyme's resistance to digestion by trypsin; presumably the modification of the enzyme surface obscured the normal points of attack of trypsin.

Torchillin et al. (1979)[13] describe the stabilization of α-chymotrypsin by the modification of surface amino groups. The 15 available amino groups were either alkylated or acylated with acrolein or succinic anhydride respectively. Modification of up to 5 amino groups with acrolein gave no significant stabilization; between 5 and 13 amino groups modified there was a marked, exponential increase in stability up to 120 times that of the native enzyme. Modification of all 15 amino groups produced an enzyme with no more stability than the native enzyme. Torchillin postulated therefore that there were just a few (2-5) amino groups whose modification might lead to stabilization. Acetylation gave a similar pattern but without such striking stabilization.

4.6 Intramolecular Crosslinking

The modification of an enzyme molecule's surface structure may be taken one stage further to the crosslinking of groups on the surface. By linking rigid molecular brackets onto and around the enzyme molecule, the potential for conformational change should, in theory, be reduced and the enzyme's stability therefore enhanced. Stabilization of oligomeric proteins may be achieved by crosslinking the subunits together. Dialdehydes, in particular glutardialdehyde, diisocyanates and bisdiazonium salts have been most commonly used as these molecular brackets. The stabilizations achieved have been somewhat variable, often being a result of single point chemical modification, largely because of the lack of any attempt to match the size of the bracket molecule to the distance between reactive groups on the enzyme's surface.

By taking the surface geometry of the enzyme into account Torchilin et al. (1977)[14] found that α-chymotrypsin was markedly stabilized when tetramethylenediamine was covalently linked to carbodiimide activated carboxyl groups on the enzyme's surface. Hexamethylenediamine, however, produced no stabilization, whilst single point attachment of aminopropanol actually destabilized the α-chymotrypsin. When the surface content of carboxylic acid groups was increased by treating the enzyme with succinic anhydride, ethylendiamine was found to be the most effective stabilizing agent. In addition the stability was even greater than for the native, tetramethylenediamine treated enzyme. Presumably the distance between modified carboxylic acid groups was decreased, requiring a shorter bracket for optimal stabilization, and the number increased allowing more brackets to be applied (Table 2).

4.7 Immobilization

If ever there was to be a goose with a golden egg in enzyme technology, immobilization of enzymes must be the number one candidate. However, there is little general agreement about the

TABLE 2. Effect of alkyl chain length of diaminoalkyl crosslinking reagent on the stability of native and succinylated α-chymotrypsin. Stability is inversely related to thermodeactivation rate constant (From ref. 14).

Chain length of cross linking alkyldiamine	Rate constants of thermodeactivation/min^{-1}	
	Native Chymotrypsin	Succinylchymotrypsin
Non-crosslinked	0.25	0.25
0	0.48	0.05
2	0.10	0.01
4	0.08	0.04
5	0.15	0.05
6	0.24	0.09
12	0.27	0.07

precise date of ovulation. Nevertheless, one of the most frequent claims for the benefits of immobilization is the potential for enhancement of the enzyme's stability. As we shall see, however, the field is positively mined with artifacts and the wise tread warily.

Immobilization has been defined as the imprisonment of an enzyme molecule in a distinct phase that allows exchange with, but is separated from, the bulk phase in which substrate, effector or inhibitor molecules are dispersed and monitored [3]. The enzyme phase is usually insoluble in water and is often a high molecular weight, hydrophilic polymer (e.g. cellulose or polyacrylamide). The imprisonment of the enzyme may be achieved by a number of means; it may be covalently bound to, absorbed to or physically entrapped within the enzyme phase.

In circumstances where the enzyme phase is comprised of a polymer matrix, the polymer may be in a defined physical state, e.g. a powder or as a membrane. The traditional approach to enzyme immobilization, that of covalently bonding the enzyme to polymer matrix, usually provides secure bonding via a great variety of reactive groups, but too often results in enzyme inactivation. Electrostatic enzyme-polymer bonding is easy to accomplish, with little inactivation caused by the mild reaction conditions, but results in a relatively insecure bonding. Copolymerization with an 'inert' protein can result in preparations with high specific activity, although inactivation of the enzyme is often observed. Entrapment of enzyme molecules inside a polymer matrix offers an easy method requiring mild

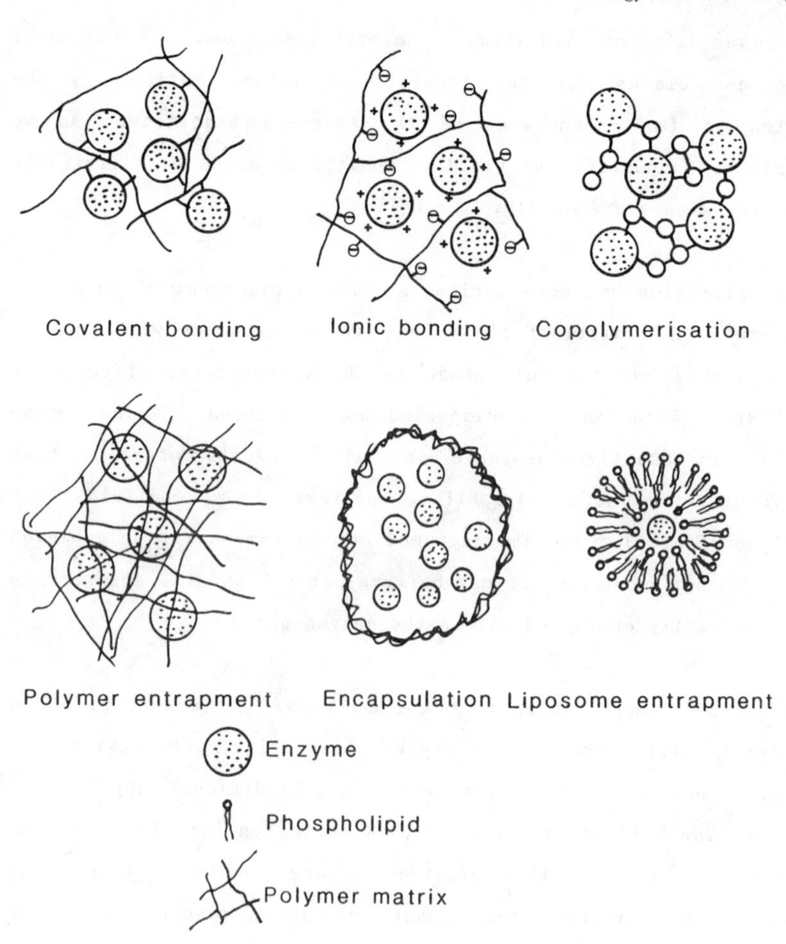

Fig.3 ***Modes of enzyme immobilization***

reaction conditions; the entrapped enzyme, however, often has its activity severely limited by diffusional restrictions and has a tendency to be leached from the polymer matrix. Encapsulation in fibres and microcapsules is a useful technique when complete separation of the enzyme from the bulk phase is required. Liposomal entrapment and other hydrophobic enzyme-'polymer' interactions are of specialist application (Fig. 3).

Immobilization may affect an enzyme's activity in a number of ways. The process of immobilization may in itself alter the <u>intrinsic</u> characteristics of the enzyme. Partitioning effects of the polymer may separately give rise to a change in the enzyme's <u>inherent</u> parameters and, in addition, diffusional constraints imposed by the polymer matrix may alter the inherent parameters to give the experimentally observed <u>effective</u> parameters.

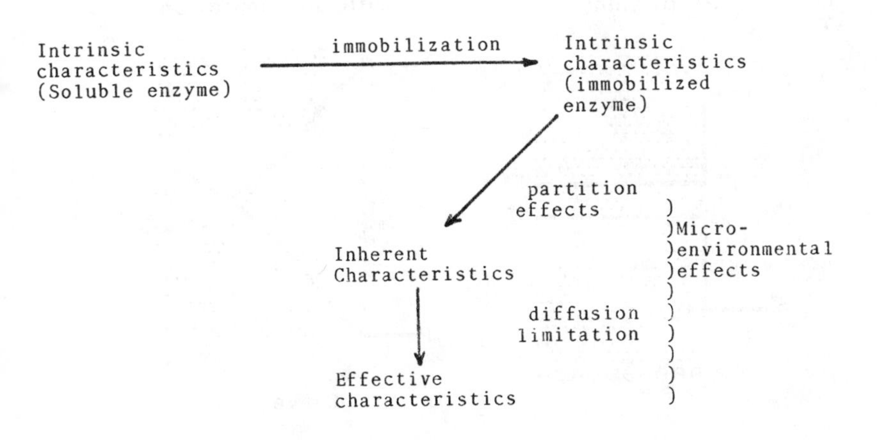

The most obvious alteration of intrinsic characteristics would be brought about by total or partial inactivation of the enzyme, caused by gross conformational changes or reaction of some essential group at the enzyme's active site. More subtle conformational changes induced in the enzyme could cause destabilization or alteration of allosteric effects or kinetic parameters. Some enzymes demonstrate enhancement of effective stability when immobilized, the reasons for which are discussed below.

The microenvironment provided by the polymer matrix affects the activity of the immobilized enzymes by causing heterogeneity of distribution of solute. Microenvironmental effects may be subdivided into partioning effects and diffusion limitation. (Fig. 4).

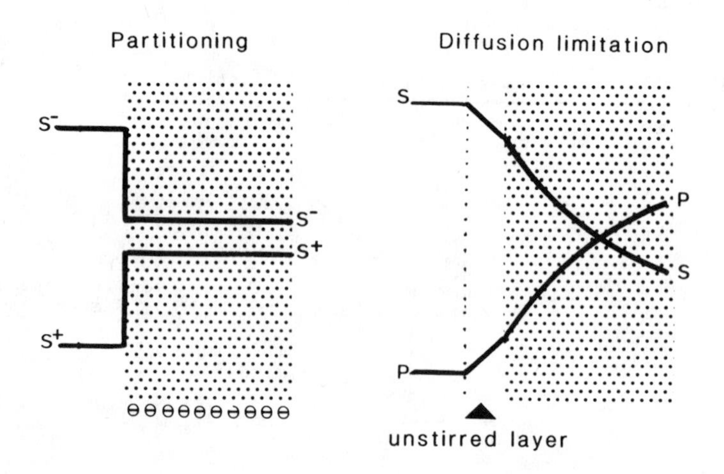

Fig.4 ***The effect of partitioning and diffusion limitation on solute concentration profiles in immobilized enzymes***

Partioning of solutes by the polymer matrix is a result of hydrophobic or electrostatic interactions and affects reacting and non-reacting solutes alike.

Limitation to free diffusion of solute molecules by the physical presence of the polymer matrix will alter the microenvironmental concentrations of solutes taking part in the reaction. The interaction between diffusion and reaction produces non-linear solute distribution throughout the polymer matrix. Diffusion limitation may be either external (i.e. up to the enzyme-polymer surface through the unstirred Nernst layer) or internal (i.e. diffusion within the polymer matrix). Where external diffusion limitation is present reaction occurs after diffusion whereas when internal diffusion limitations are present diffusion and reaction occur concurrently.

Thus immobilization provides a number of scenarios by which an enzyme may be stabilized, and many claims have been made for enzyme stabilization as a consequence of immobilization. With systems of such complexity two questions arise. Are these stabilization effects real or merely apparent? Does it matter?

To take the second question first the answer is, in many instances, probably not. Where an immobilized enzyme is to be used as, for example, an industrial catalyst or as a biosensor then the important consideration is operational stability of the whole immobilized enzyme and the cause and reality of the stability are largely irrelevant. However, if the objective is to understand how immobilization may stabilize an enzyme then the real must be disentangled from the artefactual. The rest of this discussion then will concern itself with such distinctions, starting with real intrinsic stabilization.

In retrospect, and in view of work of the nature discussed above on chemical modification and intramolecular crosslinking of enzymes, it now seems self evident that multiple point binding of an enzyme molecule to a polymer surface might bring about some degree of intrinsic stabilization. The emphasis here is on multipoint binding; obviously single point binding is unlikely to constrain conformational changes in the protein. Most usually the binding will be covalent, although electrostatic interactions may be involved. For example, trypsin attached to ethylene maleic anhydride copolymer is active in 8 mol dm^3 urea[15]; α-chymotrypsin entrapped in 50% polymethacrylic acid has a theoretical half life at 60° of several million years![16] However, unguarded optimism for this approach must be tempered by four cautions. First, the surfaces of the enzyme and polymer are unlikely to be complementary, and unless the polymer is built up around the enzyme only part of the enzyme's surface may be rigified with little consequent stabilization. Second, stabilization might be due to chemical modification of the enzyme. Third, the enzyme may be destabilized because its conformation is locked in an unstable state. Fourth, conformational change is important to the catalytic function of an enzyme, and restricting such changes may well cause inactivation or alter its kinetic characteristics (e.g. substrate specificity).

The second way in which real stabilization of an immobilized enzyme may be achieved is through the effect of protein concentration on stability discussed above. Obviously, when an enzyme is immobilized, the local concentration of enzyme within or around the polymer matrix is far greater than the average value throughout the reaction medium.

We can now turn to those stabilizing effects of immobilization which are more apparent than real: that is the intrinsic stability of the enzyme is not altered, but the effective stability of the immobilized enzyme, when compared to the free native enzyme, is enhanced. One such artefact, the differential denaturation of multiple forms of an enzyme with varying stabilities as a result of the immobilization procedure, has already been discussed (section 3).

The presence of a microenvironment around an immobilized enzyme may prevent access by denaturing agents (*e.g.* H^+ ions, organic solvents) to the enzyme molecule. For example, denaturation at unfavourable pH may be prevented by coupling an enzyme to a polyelectrolyte. The buffering capacity of the polymer in the locality of the enzyme will be far greater than could be achieved with a solution of buffer. If a polycation is used the positively charged polymer will partition H^+ ions away from the enzyme. In either case the pH around the enzyme will be different, and possibly less harmful to the enzyme, than that in the bulk phase. Dixon *et al.* (1973)[17] have reported such stabilization of lactate dehydrogenase (E.C.1.1.1.27) covalently attached to porous glass. Takahashi *et al.* (1984)[18] have reported the stabilization of catalase to the presence of organic solvents by immobilizing it onto soluble high molecular weight polyethyleneglycol. Presumably, the polyhydroxy nature of the polymer either excludes organic solvent from the enzyme's microenvironment or itself replaces the enzyme's hydration shell. Clostridial hydrogenase has been stabilized to oxygen deactivation by immobilizing it to polyethyleneimine cellulose, when it had a half-life of 1 week in air saturated water compared with 4 minutes for the free enzyme (Klibanov et al. (1978))[19]. Oxygen solubility in water is reduced as ionic strength is increased, and it seems likely that oxygen would be virtually insoluble in the highly ionic microenvironment provided by the polymer matrix. Proteolytic enzyme molecules constrained within or by a polymer matrix are unlikely to come into contact with each other and thus immobilized proteases rarely suffer autolysis.

Perhaps the most common cause of stabilization by immobilization is diffusion limitation. As we have seen the polymer matrix may present a barrier to the free diffusion of substrate molecules to the enzyme. In effect the conversion of substrate to product becomes a sequence of events, diffusion of substrate up to and into the polymer, catalysis, and diffusion of product back into the bulk phase. The overall rate will be governed by the slowest step which may be substrate diffusion. This is illustrated by Fig. 5 which schematically shows the relationship between the diffusion rate of substrate, the potential intrinsic rate of reaction and the actual effective rate of reaction of an immobilized enzyme when the substrate concentration in the bulk phase is varied. At substrate concentrations below (A) the rate is controlled by the diffusion rate, whilst above (A) it is under the kinetic control of the enzyme reaction. Clearly a higher enzyme concentration would lead to an increase in the value of A and result in the actual effective reaction rate being a function of diffusion rate for most substrate concentrations. The converse will be true for a reduced enzyme activity. It can also be clearly seen that raising the substrate concentration will overcome diffusion limitation of the effective rate of reaction. Thus if substrate diffusion is limiting, events which might cause a change in the intrinsic

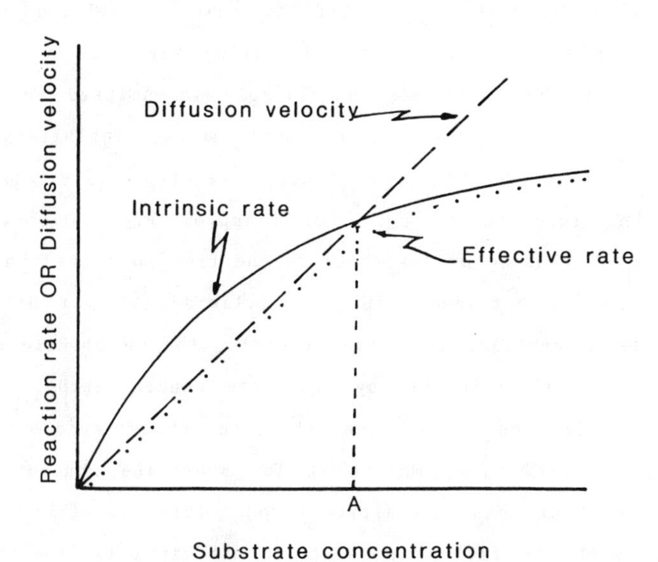

Fig.5 ***Relationship between intrinsic reaction rate(——), effective reaction rate(······) and diffusion velocity(– – –) for an immobilised enzyme***

activity of the enzyme, e.g. partial thermal denaturation or changes in pH, will have little or no effect upon the effective catalytic activity of the immobilized enzyme particle, until such time that the intrinsic enzyme activity falls below the rate of substrate diffusion and itself becomes rate limiting. To put it another way, if the polymer matrix is highly loaded with enzyme, all the substrate molecules diffusing into the particle may be converted to product by the enzyme contained within the outer shell of the polymer particle. The particle therefore has a large reserve of unused enzyme to replace any enzyme molecules inactivated by heat, pH et cetera. It is interesting to note at this point, that immobilized enzymes subject to severe diffusional limitation may not only appear very stable, but will also be largely insensitive to normal effectors of their kinetic characteristics; inhibitors will appear ineffective or the effective reaction rate may show almost no response to pH. For example, Fig. 6 shows the response of glucose oxidase both in the free form (6a) and when immobilized by entrapment in polyacrylamide (6b) at different substrate concentrations. Quite clearly, the pH profile of the free enzyme is unaffected by substrate concentration, whereas the immobilized enzyme at high substrate concentrations behaves like the free enzyme, whilst at low substrate concentrations becomes more or less insensitive to pH. This is, of course, in keeping with the fact that substrate diffusion limitation will become more pronounced as the substrate concentration is lowered. Just about the only thing which will affect the rate of reaction will be the substrate concentration and this in a linear fashion over several orders of magnitude. (Were it not

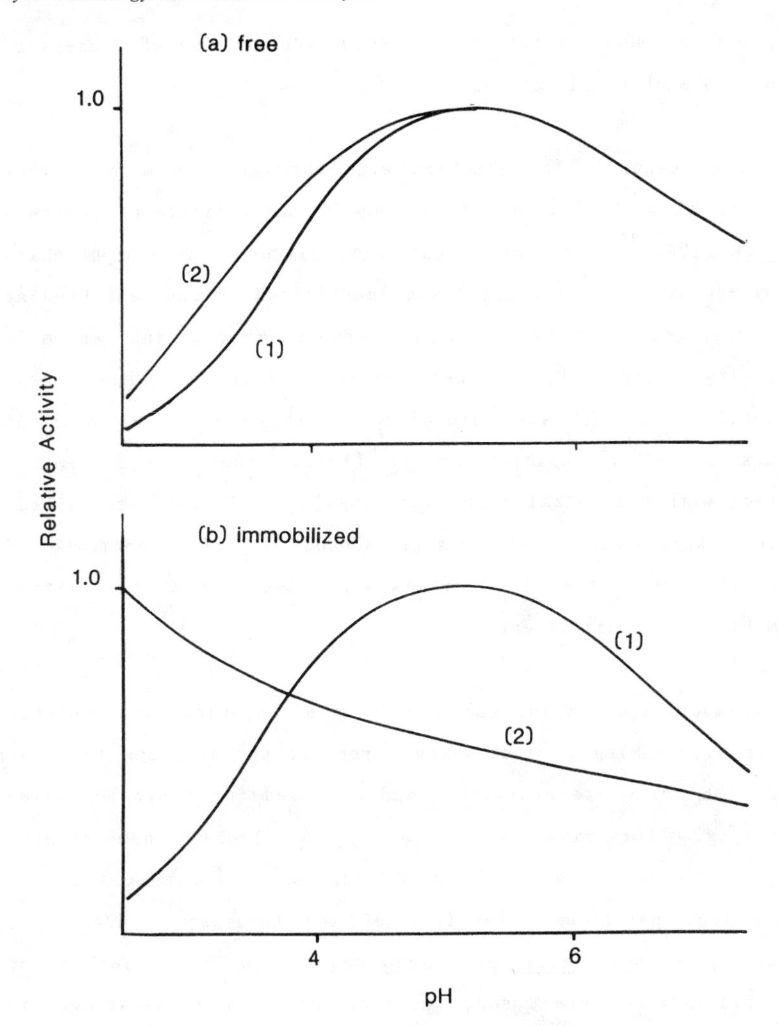

Fig.6 ***pH profiles free (a) and immobilized (b) glucose oxidase at 1.0M (1) and 0.1M (2) glucose concentrations***

for effects such as this the construction and use of biosensors would be much complicated).

We shall conclude this section with three of the more curious effects of stabilization of enzymes by immobilization. Carrera et al. (1982)[20] discovered that when glyceraldehyde-3-phosphate dehydrogenase (E.C.1.2.1.12) was immobilized by covalent bonding to cyanogen bromide activated sepharose 4B, the enzyme's activity increased by 2-3 times within an hour from immobilization and was maintained at this level for over 30 hours at 40°C. Cashion et al. (1982)[21] describe a similar effect with calf alkaline phosphatase (E.C.3.1.3.1) immobilized onto trityl-agarose; the enzyme showed a gradual increase of activity which after 12 days was 4-8 times that of the initial immobilized or native enzymes.

Cell immobilization is, for a variety of technical and economic reasons, becoming an increasingly popular pursuit; for the most part the cells are non-viable and are used for a single enzyme activity. For example, Takata et al. (1982)[22] demonstrated that different methods of immobilization had a marked effect upon the operational stability of the fumarase (E.C.4.2.1.2) activity of Brevibacterium flavum cells. Two basic methods of immobilization were used, polyacrylamide and K-carrageenan entrapment both with and without the addition of polyethyleneimine. The various preparations could be ranked in order of increasing stability, free cells, free cells plus polyethyleneinine/polyacrylamide entrapped cells, polyacrylamide entrapped cells plus polyethyleneimine, K-carrageenan entrapped cells, and most stable K-carrageenan entrapped cells treated with polyethyleneimine. This latter preparation was some 22 times more stable than the polyacrylamide preparation. They concluded that this high stability towards heat, urea, high pH and ethanol (see Table 3) was a result of a three way interaction between the cell membrane, polyethyleneimine and the K-carrageenan. They did not speculate how such an interaction could stabilize an individual enzyme contained with the cell, nor is it easy to explain. It is possible that some of the stabilization effect could have been real and due to the presence of the polyethyleneinine, as polyamines are known to stabilize cells and are used as cryoprotective agents. However much of the observed stabilization could have been due to diffusion limitation. Certain indicators of diffusion limitation were present, for example broadening of pH profiles, less than expected increase of catalytic activity with increases in temperature, high catalyst loading. The differences observed between the different immobilization matrices could simply reflect variations in polymer structure and hence diffusional resistance; the inclusion of polyethyleneimine changes some of the physical characteristics of the K-carrageenan gel. In addition, the initial loading of cells/enzyme in the preparations could have been affected by the immobilization method. Acrylamide monomers and the free radicals produced by the polymerization reaction in the formation of polyacrylamide gels are known to be toxic and it is thus likely that these preparations contained a lower active concentration of cells/enzyme than the K-carrageenan.

TABLE 3. Comparison effect of immobilization method on stability of fumarase of *Brevibacterium flavum* (From ref. 20)

Cells and Immobilization Method	Initial fumarase activity(units/ml gel) at 37°C	Half-life (days) at 37°C	Relative remaining activity(%) after heat treatment	
			@ 55°C for 60 min	@ 60°C for 15 min
Free Cells				
Native	n.d.	n.d.	18	0
+0.15% polyethyleneimine	n.d.	n.d.	69	0
Immobilized cells				
Polyacrylamide	10.2	94	60	0
Polyacrylamide (B.ammoniagenes)	8.8	53	n.d.	n.d.
Polyacrylamide +polyethyleneimine	n.d	n.d.	88	37
K-carrageenan	15.0	160	100	59
K-carrageenan +polyethyleneimine	16.3	243	100	100

5. CONCLUSION

We have seen in the course of this discussion how enzyme stability may be enhanced and why this is desirable. Many of these methods may indeed yield enzyme preparations of enhanced intrinsic stability, albeit at some cost, while for others, notably immobilization, enhanced stability may be apparent rather than real. To confuse the picture enzyme deactivation is not necessarily a simple monomolecular process but may be a complex event. It should be clear, therefore, that great caution must be exercised when claiming superior intrinsic stability for an enzyme, as real stabilization as a result of manipulating the enzyme is probably the exception.

In practice, however, operational stability of a biocatalyst is the goal and the exact cause or explanation of enhanced stability is of secondary importance. He cares not for the means who profits by the ends.

REFERENCES

1. S.L. Neidleman. In Biotechnology and Genetic Engineering Reviews. Vol 1 (1984). ed. G.E. Russell. Intercept, Newcastle. U.K. pp. 1-38.

2. A.M. Klibanov. Science (1983). 219, 722-727.

3. M.D. Trevan. "Immobilized enzymes: Introduction and Application in Biotechnology". (1980). John Wiley, Chichester, U.K.

4. V.V. Mozhaev and K. Martinek. Enzyme Microb. Technol. (1984). 6, 50-59.

5. J.P. Henley and A. Sandana. Enzyme Microb. Technol. (1985). 7, 50-60.

6. R.A. Beardslee and J.C. Zahnley. Arch Biochem. Biophys. (1973). 158, 806-811.

7. R.D. Schmidt. Adv. in Biochem. Eng. (1979). 12, 41-118.

8. A. Zaks and A.M. Klibanov. Science. (1984). 224, 1249-1251.

9. G. Forniani., G. Leoncini., P. Segni., G.A. Calabria and M. Dacha. Eur.J.Biochem. (1969). 7, 214-222.

10. C. Tanford. Science. (1978). 200, 1012.

11. D.J. Merkler., C.K. Farrington and F.C. Wedler. Int.J.Pept.Protein Res. (1981). 18, 430-437.

12. P. Tuengler and G. Pfleiderer. Biochim.Biophys.Acta. (1977). 484, 1-8.

13. V.P. Torchillin., A.V. Maksimenko., I.V. Berezin., A.M. Klibanov and K. Martinek. Biochim.Biophys.Acta. (1979). 567, 1-11.

14. V.P. Torchillin., A.V. Maksimento., V.N. Smirnov., I.V. Berezin., A.M. Klibanov and K. Martinek. Biochim.Biophys.Acta. (1977). 522, 277-283.

15. L. Goldstein. Biochemistry. (1964). 3, 1913-21.

16. K. Martinek., A.M. Klibanov., A.V. Tchernysheva., V.V. Mozhaev., I.V. Berezin and B.O. Glotov. Biochim.Biophys.Acta. (1977). 485, 13-28.

17. J.E. Dixon., F.E. Stolzenback., J.A. Berenson and N.O. Kaplan. Biochim.Biophys. Res.Commun. (1973). 52, 905-912.

18. K. Takahashi., H. Nishimura., T. Yashimoto., M. Okada., A. Ajima., A. Matsushima., Y. Saito and Y. Inada. Biotechnol.Letts. (1984). 6, 765-770.

19. A.M. Klibanov., N.O. Nathan and M.D. Kamen. Proc.Nat.Acad.Sci. U.S.A. (1978). 75, 3640-3643.

20. G. Carrea., R. Bovara and P. Pasta. Biotechnol. Bioeng. (1982). 24, 1-7.

21. P. Cashion., A. Javed., D. Harrison., J. Seeley., V. Lentini and G. Sathe. Biotechnol. Bioeng. (1982). 24, 403-423.

22. I. Takata., K. Kayoshima., T. Tosa and I. Chibata. J.Ferment.Technol. (1982). 60, 431-437.

11

Applications of Biotechnology to Chemical Production

By D. J. Best

BIOTECHNOLOGY CENTRE, CRANFIELD INSTITUTE OF TECHNOLOGY, CRANFIELD, BEDS. MK43 0AL, U.K.

1. Introduction

The prime importance of biotechnology in the production of chemicals is that it provides an additional complement of catalysts to the armoury of the synthetic chemist. Although biological catalysts possess unique advantages and disadvantages, as catalytic systems they are no different to conventional catalysts; they provide the means whereby a given reaction, or set of reactions, is performed at a much faster rate, and often with a much greater degree of specificity, by lowering the activation energy of that reaction(s). Traditionally, biological methods of catalysis have been used independently of chemical processes and are, consequently, less well characterised as catalytic systems. The emergence of 'new biotechnology', through the advances in genetic and protein engineering, has lead to a fundamental reassesment of biological catalysts, divested of their traditional mystique, as catalytic units and their use, in combination with chemical catalysis, to evolve pathways and processes that, until now, have not been feasible.

However, in order to assess the potential of biocatalysis in replacing or complementing chemical catalysis, it is necessary to delineate the characteristics and limitations of these systems. On an industrial scale the catalytic process, whether chemical or biochemical, has to be viable economically. Biocatalytic processes, however novel or unique, have to be judged finally on this criterion. Of course, the term 'biotechnology' is merely a wider umbrella term for what was traditionally referred to as 'applied microbiology' and now encompasses a plethora of related disciplines with the aim of commercialising biological processes. The food industry has been the home of this technology for many thousands of years, with the production of beers, wine, cheese and other fermented products. The application of many biocatalytic processes is still to be found in this industry, although this century has witnessed the use of biologically-based processes in the chemical (e.g. solvent production before the Second World War) and pharmaceutical industries (e.g. antibiotic and steroid production). This chapter describes the essential properties of biocatalytic systems, their advantages and disadvantages, and illustrates these with specific examples of direct applications of biotechnology to chemical production.

2. Characterisation of biotechnological processes for chemical production

2.1. Reaction sequence.

A biocatalytic process can involve either a single reaction step, catalysed by one enzyme, or a multi-step process, where a series of enzymes, constituting either part or an entire metabolic pathway, is employed to generate the desired chemical product. When the process is a multi-step procedure it is usually performed without the isolation of the participating enzymes: i.e. they are utilised in situ within the organism. The organism may, in this instance, be regarded as a 'bag' of catalytic units arranged in the correct process orientation.

Single step reactions frequently utilise isolated enzymes as catalytic units, although the alternative exists to again use the enzyme in situ in the organism, if the particular biocatalytic unit is either too costly to isolate or too unstable in its isolated form. The disadvantages of using single enzymes in non-isolated form are three-fold (see also Figure 4):

(i) Restricted access of the substrate, altering the reaction kinetics. In certain cases, this may be overcome by permeabilisation of the cell membrane by chemical treatment.

(ii) The possible removal of either substrate or product by other enzymes or enzyme pathways in the host organism.

(iii) The possible contamination of the product due to cell lysis.

A number of single step biocatalytic reactions are extensively used in industry for the generation of chemical products and these reactions cover a number of

TABLE 1: Single step reactions of commercial importance

Reaction Type	Catalyst	Reaction Catalysed	Location	Use
Isomerisation	Glucose isomerase (EC 5.3.1.5)	D-glucose → D-fructose	Isolated or *in situ*	Production of high fructose syrups
Hydrolysis	*alpha*-amylase (EC 3.2.1.1)	Starch → glucose + maltose	Immobilised	Production of high DE syrups
Hydrolysis	Lactase (EC 3.2.1.23)	Lactose → glucose + galactose	Isolated and immobilised	Rectification of whey permeate
Hydrolysis	Lipase (EC 3.1.1.3)	Fats → fatty acids + glycerol	Isolated	Improvement of quality of foods
Hydrolysis (Condensation)	Penicillin acylase (EC 3.5.1.14)	Side chain cleavage (reverse, condensation)	Immobilised	Production of semi-synthetic penicillins
Condensation	Aspartase (EC 4.3.1.1)	Fumarate + ammonia → aspartate		Amino acid production
Hydroxylation	Tyrosinase (EC 1.14.18.1)	Tyrosine + water → L-DOPA + oxygen		Production of antihypertensives

Table 1 (continued)

Reaction Type	Catalyst	Reaction Catalysed	Location	Use
Oxygenation	Undefined oxygenative systems	Stereospecific steroid hydroxylation	Whole organism	Steroid transformations
Peroxidation	Lipoxygenase (EC 1.13.11.12)	Carotenoids + linoleic acid + oxygen ⟶ peroxidised linoleic acid		Peroxidised oils used in flavourings
Oxidation/ reduction	Undefined enzyme systems	Stereo- and regio-specific oxidations and reductions	Whole organism	Steroid transformations

reaction types (Table 1).

Multi-step biocatalytic reactions have been utilised for many years in the brewing and dairy industries[1]. Partial or total metabolic pathways are utilised for the synthesis of a particular chemical product from plentiful and cheap feedstocks, such as starch and molasses. Two types of metabolic pathway are involved in chemical production. Anabolic pathways are used to convert a simple substrate, such as glucose, to much more complex chemical products, such as antibiotics, vitamins, terpenes, and biopolymers. Catabolic pathways break down the feedstock into much more simple chemicals, such as ethanol, acetone and butanol (Table 2). These processes are usually associated with the growth of the producer organism; i.e. the desired chemical production is an end result of the growth of the organism upon a particular substrate. This may be termed growth-associated biocatalysis and the production of ethanol by fermentation is an example of this type of process[2]. The regulation of many pathways involved in the biological production of metabolites is still poorly understood and the actual production period may occur after growth has, to a large extent, ceased. This may be termed post-growth biocatalysis and can be logically extended to cover the use of multi-step pathways in non-growing organisms that are, for instance, immobilised or used for the transformation of substrates upon which they cannot grow. In addition, the generation of the desired product from these anabolic and catabolic pathways may necessitate the manipulation of the appropriate metabolic sequences, to divert carbon flow into product formation/accumulation. Thus the normal metabolic control processes within the organism must be disrupted and any repression of the activity of the target partial metabolic pathway, that might occur during such unbalanced product synthesis and accumulation, overcome. There are three ways in which manipulation of metabolic pathways has been achieved to effect the desired multi-step chemical conversion (Figure 1):

- alteration of environmental conditions, under which the organism is grown, to achieve a metabolic imbalance (e.g. organic acid synthesis[3]).
- disruption of the pathway by the use of substrate or product analogues

TABLE 2: Multi-step reactions of commercial importance

Category	Example
Anabolic	Amino acids
	Antibiotics
	Carotenoids
	Vitamins
	Biopolymers
Catabolic	Solvent fermentations
	:ethanol
	:acetone/butanol
	:butanediol
	:glycerol
	Organic acids
	:citric
	:acetic
	:gluconic

FIGURE 1: Manipulation of microbial metabolism for chemical production

(a) Environmental parameters

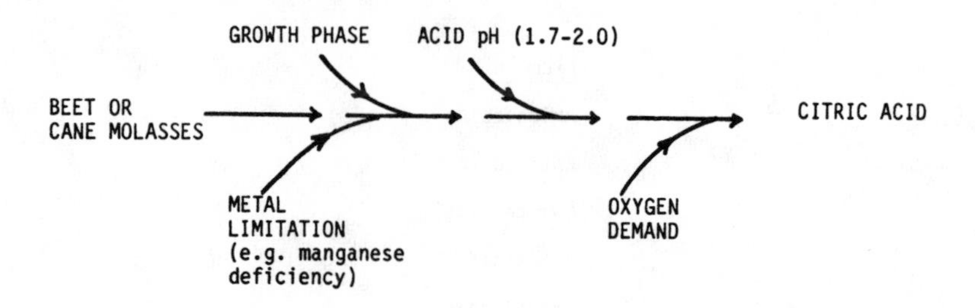

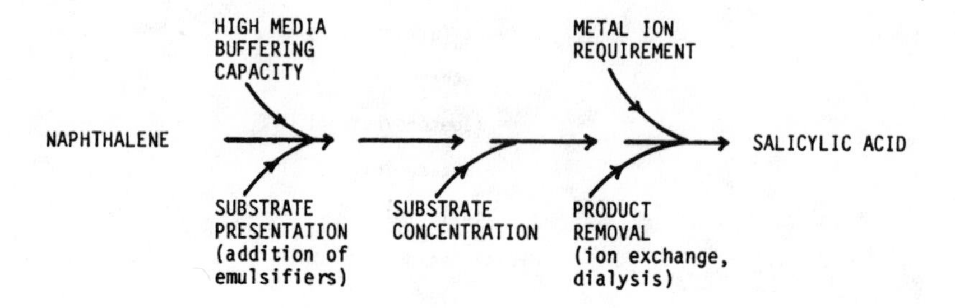

(b) Directed synthesis

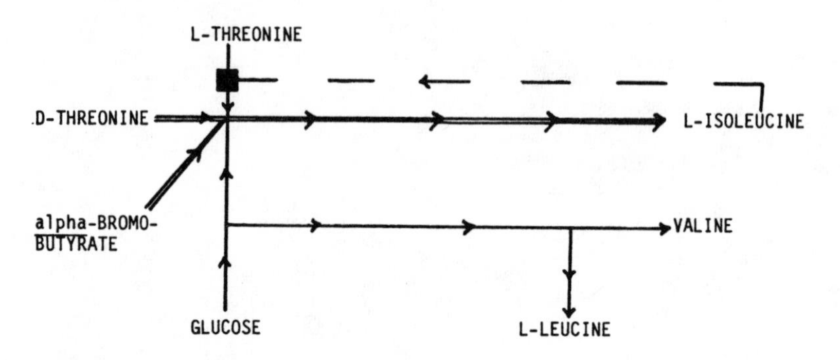

■ : site of enzymatic inhibition
═ : directed metabolic route

FIGURE 1 (continued)

(c) Mutant generation

(i) Chemical mutagenesis

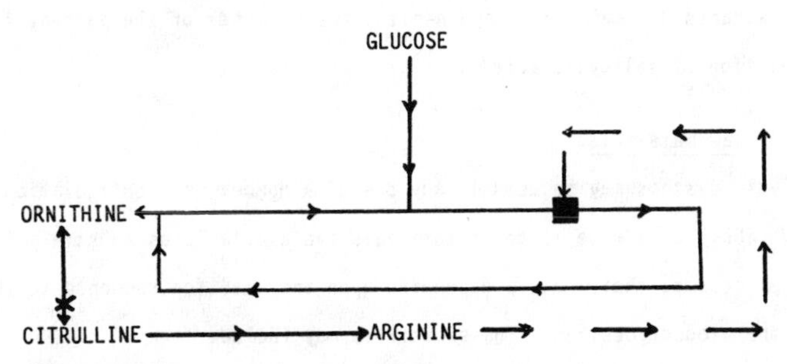

■ : site of negative feedback inhibition
✕ : key enzymic lesion

Auxotrophic mutants are unable to produce negative feedback inhibition of a particular pathway due to the absence of a key enzyme.
Regulatory mutants have lost some degree of regulative control of biosynthesis and are therefore not sensitive to such feedback inhibition.

(ii) Genetic engineering

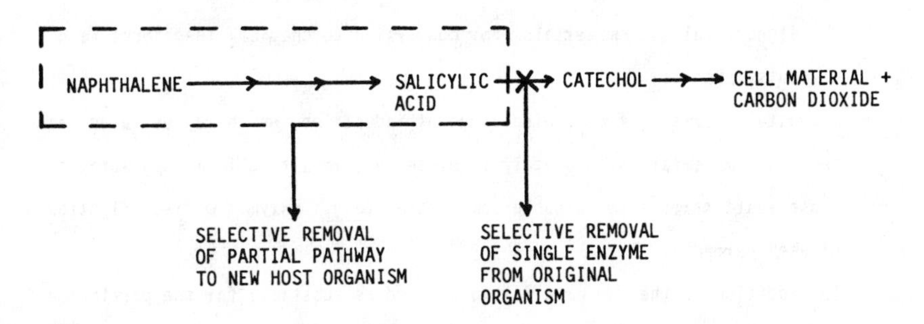

✕ :site of required enzymic lesion

to specifically inhibit or interrupt given enzyme sequences or bypass metabolic control points (directed synthesis[4]).

- generation of mutants of the producer organism, either by chemical mutagenesis or by advanced genetic engineering, so that only partially utilised metabolic sequences are constructed within the organism (e.g. blocked mutants in amino acid synthesis[5], the transfer of the pathway for the generation of salicylic acid[6]).

2.2. Source of raw materials.

Biocatalytic systems may be used in any one of a number of configurations, as described above. A wide range of materials are available as substrate for biocatalytic processes, their nature depending upon the reaction sequence to be utilised and the product desired. The substrates may include:

- purified, speciality substrates that will undergo simple transformations in single or limited multi-step reactions (e.g. amino acids, triglycerides, menthol esters).
- carbohydrate feedstocks (starches, molasses) for interconversion to more simple (e.g. ethanol, organic acids) or more complex (e.g. xanthan gums) products.
- commodity chemicals, such as methane, methanol or propene.
- lignocellulosic materials for conversion to chemical feedstocks (e.g. solvents, phenols).
- waste materials, from a wide range of industries, which may be used as the raw material for biocatalytic processes, resulting in an upgrading of those waste streams to valuable commodities (e.g. enzymatic rectification of whey permeate).

In addition to the raw material to be used as substrate for the particular chemical conversion, it is also necessary to consider the substrate to be used for the generation of the biological catalyst itself. This is particularly important when the biocatalytic process is not growth associated and the active catalyst has to be generated prior to use in the process (see Section 2.6, 'Biocatalyst stability').

2.3 Reaction specificity.

The specificity of a reaction or reaction sequence is one of the main advantages of biocatalysis over conventional catalysis. The specificity is determined by the capability of the catalyst molecule to 'recognise' the substrate molecule, in many instances with both regiospecificity (which atoms or functional groups of the substrate molecule will participate) and stereospecificity (the orientation of attack at a particular site). Both these properties are determined by the extent to which the active catalytic site on the enzyme molecule can recognise the substrate by molecular interaction (e.g. regiospecificity of xanthine oxidase[7], the enzymatic resolution of racemic mixtures[8,9], site specific oxyfunctionalisation of steroids[10]: Figure 2a-c). Not all enzymes are as specific in their mode of action and substrate specificity as others. This diversity of specificity may be related to enzyme mechanism and may be used to advantage in some instances in controlling metabolic pathways, as has been described in the use of substrate analogues[11].

For enzymes requiring two substrates for a particular reaction (e.g. electron donor and electron acceptor), specificity for either one of the substrate pair may be quite different (Figure 2d). The single substrate enzyme glucose isomerase is fairly specific for its substrate, glucose, and this is, to some extent, dependent on the microbial source of the enzyme[12,13]. Glucose oxidase, routinely used in clinical diagnostics for glucose determinations, is specific for the carbohydrate substrate but not the electron donor, oxygen. Oxygen may be replaced by benzoquinone and the enzyme reaction reduces this to hydroquinone[7]. Alkane monooxygenases, on the other hand, which insert single oxygen atoms into chemically inert alkanes (such as methane), can be relatively catholic in the range of substrate oxygenated[14] but are very specific for the electron donors, reduced pyridine nucleotides. Several classes of enzyme, such as the cytochrome P-450 oxygenases involved in detoxification, are also non-specific in the type of reaction catalysed: oxygenations, demethylations, dehalogenations and epoxidations may all be catalysed by the same protein catalyst[15].

FIGURE 2: Biocatalytic reaction specificity

(a) Stereospecificity - e.g. resolution of racemic mixtures

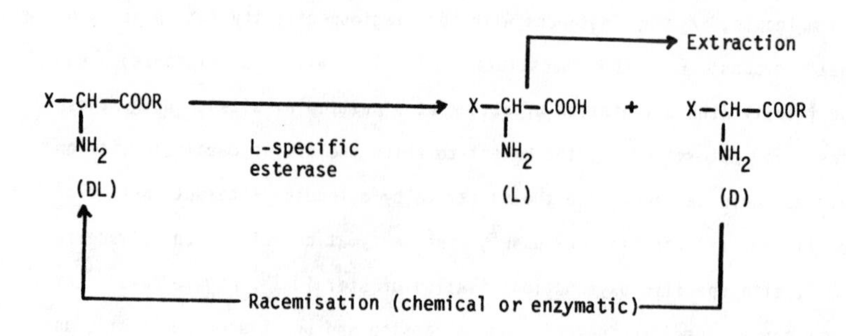

(b) Regiospecificity - e.g. effect of substituents on aromatic aldehyde oxidation

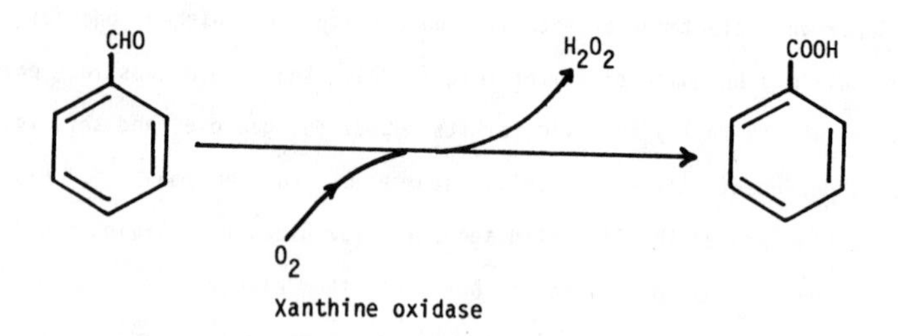

Substituent group	Rate(para)/Rate(ortho)
$-F$	0.7
$-I$	113.0
$-OCH_3$	2.0
$-OCH_2CH_2CH_3$	34.0
$-NO_2$	500.0

FIGURE 2 (continued)

(c) Stereo- and regiospecificity - e.g. the microbial hydroxylation of steroids

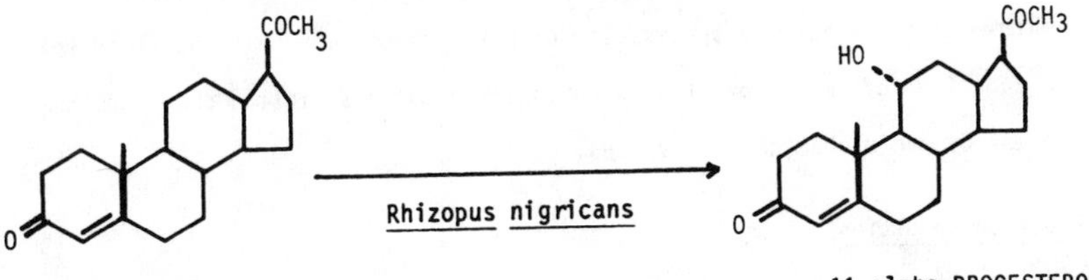

$COCH_3$ $COCH_3$

HO

O O

Rhizopus nigricans

PROGESTERONE 11-alpha-PROGESTERONE

(d) Substrate specificity

ENZYME	REACTIONS CATALYSED
Glucose isomerase	Glucose → Fructose
Glucose oxidase	D-glucose + O_2 → D-gluconolactone + H_2O_2
	D-glucose + benzoquinone → D-gluconolactone + hydroquinone
Methane mono-oxygenase	CH_4 → CH_3OH
	$CH_3CH_2{=}CH_2$ → CH_3CH_2–O–CH_2 (epoxide)
	benzene → phenol (OH)

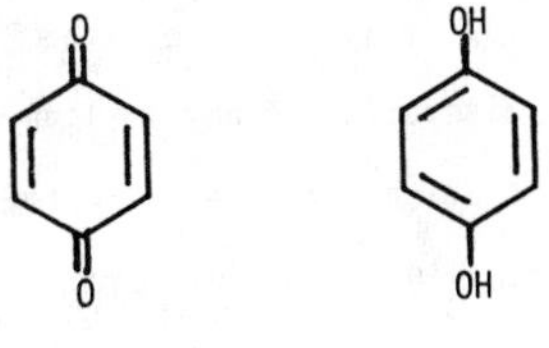

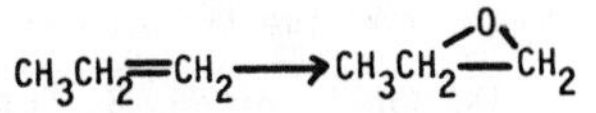

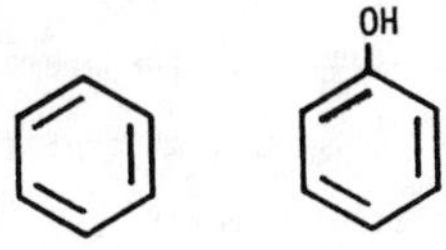

O OH

O OH

All reactions require molecular oxygen and reduced pyridine nucleotides

Therefore, it is necessary to define the specificity of the biocatalytic system, both for substrate and reaction type, as this may allow enhanced control over that system and its use for unexpected, unconventional reactions. The broad nature of the substrate specificity of many enzymes may be responsible for the degradation of many of the xenobiotic compounds introduced into the environment by man's chemical activities.

2.4 Reaction conditions

The reaction conditions under which biological catalysis occurs are usually extremely mild in comparison with those conditions required for the equivalent chemical conversion. Reactions occur over a moderate temperature range (20° - $100^{o}C$), at normal atmospheric pressures (although some fermentations may be carried out at elevated pressures to satisfy oxygen demand) in predominantly aqueous environments. For example, the biological activation and oxidation of methane to methanol occurs at $30^{o}C$ at one atmosphere pressure[16], whereas the same chemical conversion requires much more extreme reaction conditions and two steps. Whilst the mild conditions are advantageous in terms of the reduced energy input required for the desired reaction, they may be disadvantageous in terms of relatively low reaction rates and the extent to which these may be elevated by alteration of physical parameters.

The operation of these systems in aqueous environments is also problematical, in terms of processing dilute aqueous solutions for product recovery downstream of the reaction system. Therefore the reaction conditions have to be manipulated to maximise the product concentration in the product stream, until the value of the final product is such that the cost of the recovery procedure can be absorbed by the overall economics of the process. Traditional fermentative procedures have attempted to maximise the product concentration in the fermentation broth: for instance, in organic acid production levels of 80 gm/l citric acid are obtained from sucrose[17]. Patented processes for the generation of dicarboxylic acids from n-alkanes using hydrocarbon-utilising yeasts quote levels of 60 gm/litre for C_{16}-dicarboxylic acids (Figure 3)[18].

FIGURE 3: Generation of aliphatic dicarboxylic acids from the corresponding n-alkanes using microbial oxidative processes

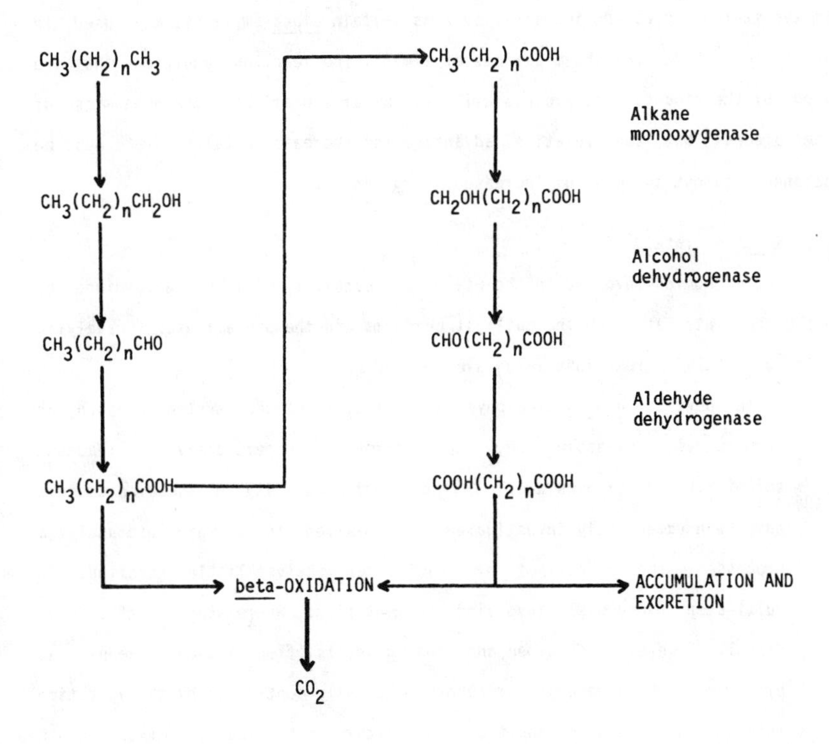

There is an increasing interest in the elaboration of thermophilic reaction systems to replace traditional mesophilic biocatalysis. Biocatalytic systems that can operate in excess of 60°C have the double advantage of increased reaction rate and, usually, increased stability. Some commercially used enzymes in the starch processing industry, such as certain alpha-amylases, are used in excess of 100°C for short periods of time[19]. The reaction conditions that are used for the biocatalytic process will have a large bearing on the economics of the process but the relative advantages of increased catalytic rate must be balanced against the cost of increased energy input.

2.5 Reaction rate.

The economics involved in the use of any catalyst will be determined by both the rate at which the catalyst performs and the overall useful catalytic life. Biocatalytic reactions rates are defined by;

- The physical and chemico-physical parameters of the system - such as temperature, pressure, pH, ionic strength, water activity, reactant solubility, vapour pressure. Although the majority of these parameters have been extensively investigated in relation to certain biocatalytic processes, the effect of pressure often receives little attention. In multi-step procedures involving fermentation, where the control of dissolved levels of oxygen and other gases is often critical, the partial pressure of these gaseous components is usually controlled by the agitation rate and the gaseous composition passing through the reaction mixture. In processes with high oxygen demands, such as waste treatment, these levels may be met by using innovative fermenter design to increase gaseous transfer at higher pressures (the deep-shaft air-lift fermenter)[20] or pure oxygen streams[21].
- the reaction rate will also be determined by the inherent parameters of the catalyst system itself. For catalysis by isolated enzyme(s), this will largely be reflected by the affinity of the enzyme for the substrate(s) (K_m), the rate of catalytic turnover (i.e. the number of catalytic events per second...this ranges from $3 \times 10^{-3} - 10^{5} s^{-1}$ at normal physiological

temperatures)[22], which will be reflected by the maximum reaction velocity (V_{max}), and the response of the catalyst to inhibitors of the reaction (either substrate, product, analogues of either or unrelated inhibitors). Alteration of any of these parameters is an attractive target for protein engineering. For processes in which a single enzyme is utilised *in situ* in the organism, or in the case of multi-step reactions, the situation is further complicated by various diffusional barriers presented to the passage of the substrate and/or product by the compartmentalisation of the catalyst (Figure 4). This problem is also often encountered in immobilised systems. Cells, of whatever biological origin, may possess mechanisms to actively transport substrates and products in and out or these materials may enter by passive diffusion. These processes will effect the rate at which the reaction will occur, due to the concentration gradients that may develop. This may necessitate manipulation of the catalyst itself, by permeabilisation of the cell to allow free diffusion, or by manipulation of environmental parameters, such as the inclusion of agents to increase the dissolved concentration of substrate or decrease the dissolved concentration of product. This is particularly relevant in the case of poorly soluble substrates. For example, the availability of hydrocarbons may be increased by the inclusion of detergents in the reaction mixture[23]. The course of a fermentation, such as that of naphthalene to salicylic acid, may be drastically improved by the addition of an ion exchange resin to remove, continuously, the acid product as it is generated[24].

The rate and extent of the reaction will also be controlled, in the case of multi-step processes, by the degree to which either substrate or product are removed by other metabolic sequences, and thus made unavailable for the desired transformation. The reactants are therefore wasted by unwanted side reactions and these may be removed by the appropriate manipulation of the system, either by chemical inhibition of the pathway or, more specifically, by rupturing or blocking a catalytic sequence by genetic engineering (*e.g.* increase in the rate of dicarboxylic acid formation by blocking the enzymes of *beta*-oxidation[18]:

FIGURE 4: Diffusional constraints imposed upon biocatalytic reactions in isolated cells, microbial aggregates and immobilised systems

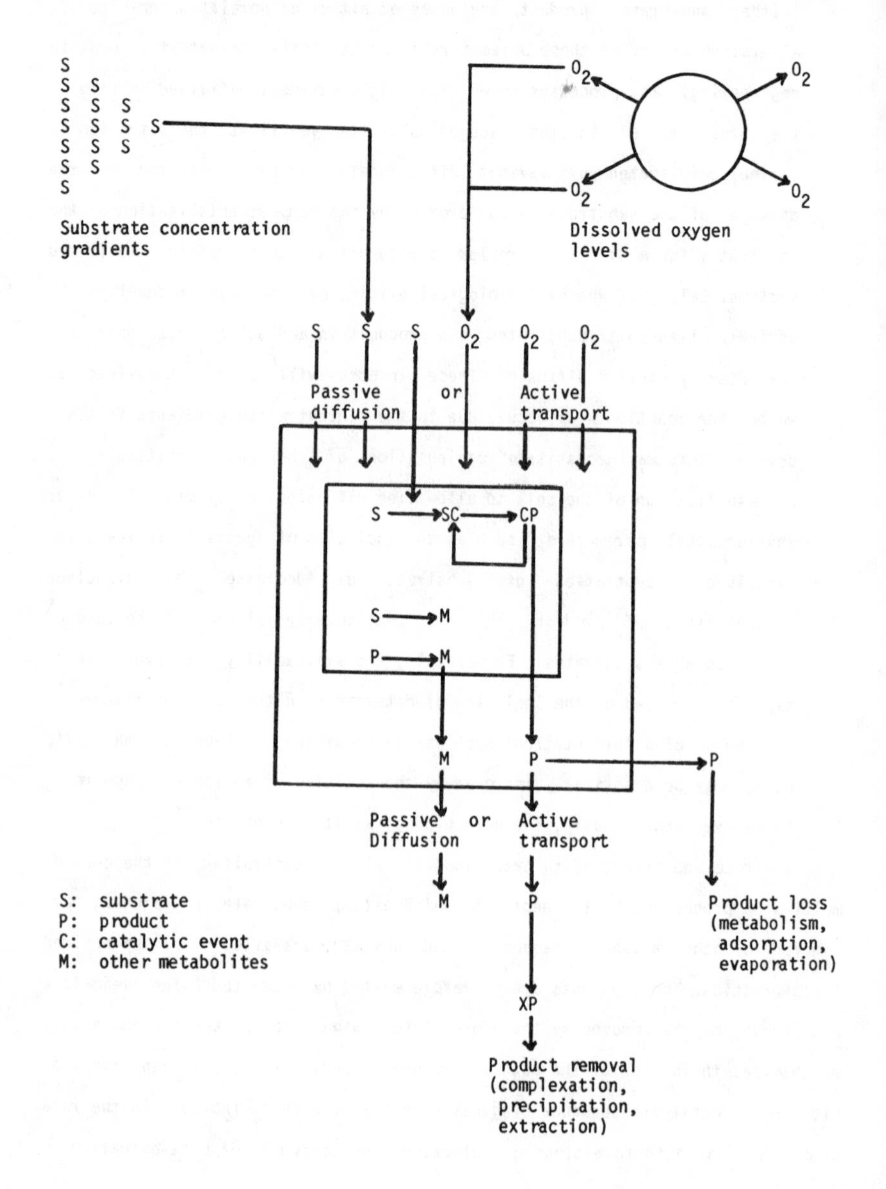

Figure 3). Alternatively, the metabolic pathway may be removed by genetic engineering to a host organism that does not possess the appropriate degradative machinery (e.g. naphthalene conversion to salicylate; Figure 1(c)). In addition, the rate limiting step of a particular metabolic process, once identified, may be subject to improvement by genetic means. In the production of single cell protein (the animal feed 'Pruteen') from methanol by Methylophilus methylotrophus, an inefficient step was identified as the incorporation of nitrogen into cellular material. A more efficient enzyme system was engineered into the organism (Figure 5), with the result that the efficiency of the overall conversion process was improved[25]. The final parameter affecting reaction rate is therefore the efficiency of conversion to the desired product. Side reactions of the catalyst itself or of other systems present should be identified and disposed of by manipulation of the system.

2.6 Biocatalyst stability.

The application of biocatalytic systems to chemical conversion is also dependent upon the stability of the catalyst itself. The reaction catalysed may be novel, fast, efficient, stereo- and regio-specific and produce a high added-value product but, if the catalyst life is of the order of minutes or hours only, the economics of regenerating the catalyst will outweigh all the other advantages and prevent implementation of the catalytic process. Stability of an enzyme system may be defined in terms of a number of parameters: (a) longevity of the catalyst with respect to inactivation by denaturation of the catalyst molecule, causing structural alterations in the active site, due to temperature, pressure, shear; (b) longevity of the enzyme with regard to degradation by specific proteases, i.e. catalyst turnover and regeneration; (c) longevity of the catalyst with respect to the number of catalytic cycles through which it turns; (d) irreversible catalyst poisoning.

The major stumbling block to the implementation of a wide variety of biocatalysts to processes of commercial interest is this lack of stability and has particularly hindered the application of the oxygenase type of catalyst to the introduction of single oxygen atoms into a wide variety of organic

FIGURE 5: Alternative routes of ammonia assimilation in bacteria. The gene for the energetically less expensive glutamate dehydrogenase pathway was introduced into *Methylophilus methylotrophus* to replace ammonia assimilation by the GS/GOGAT pathway.

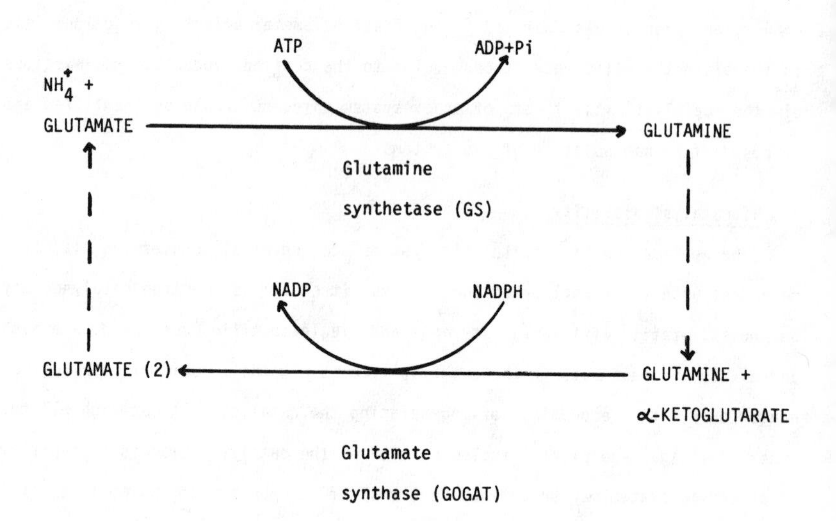

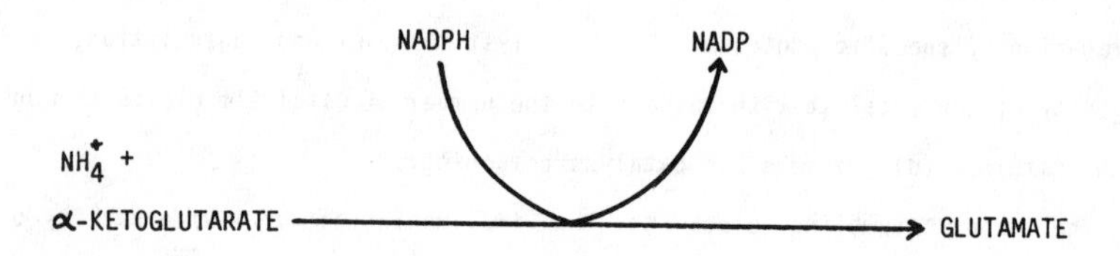

molecules. For this reason many biocatalytic systems are used in the native form and are not isolated and used individually, unless of exceptional stability. The search for thermophilic biocatalytic systems has been designed, in part, to develop improved catalyst stability. As indicated earlier, the question of stability may be intimately related to the mechanism of action and those enzymes which have to generate and control very active reactant species, such as the oxygenases, may suffer from auto-inactivation if not protected by their conventional environment. Inactivation by denaturation of the active site is of special importance.

Protein degradation and turnover of the catalyst are important if the whole cell is to be used as the catalyst. Catalyst turnover can be caused by the degradation of the entire molecule through the action of proteases. Several alternative solutions are available. The degradative machinery of the cell may be removed or selectively inactivated or the catalyst molecule itself may be continuously regenerated. If the chemical transformation process is growth-associated this does not present a major problem, as the organism/cells are synthesising protein and continuously replacing lost enzymes. However, for post-growth procedures two courses of action are open, depending upon the nature of the systems involved. Enzyme systems that are present constitutively in the organism (that is always synthesised whatever the growth substrate) may be regenerated by the addition of sufficient carbon and nitrogen to permit renewed protein synthesis. In such systems, protein synthesis may be allowed to tick over at a low rate during the biocatalytic procedure. Enzymes that are inducible demand the presence of the requisite substrate. Genetic information coding for the synthesis of a particular set of enzymes may be switched on by the presence of that inducer. If the inducer of the system is also the substrate for the process, this may present few problems but, if the inducer is not the substrate in use (for instance, a substrate analogue transformation), then some means of switching on the genetic machinery is necessary. Another possibility exists in which the desired enzyme(s) for the tranformation may be induced during growth on an economic feedstock by the inclusion, during growth, of low levels of a substrate analogue that will induce the system. It is also important to ensure that expression of the requisite enzymes is not suppressed by the growth substrate under these circumstances.

An inducible system with potential for a particular chemical transformation is an excellent target good for genetic engineering so that the control of the process may be understood and then manipulated to become constitutive. Modification of the relevant genes with powerful, high frequency promoter sequences would remove the need for an inducing substrate. A recent process used a constitutive mutant of a species of Pseudomonas putida for the primary dihydroxylation of benzene, as a preliminary step to the formation of polyphenylene[26] (Figure 6). The genes coding for naphthalene oxygenase have been tranferred and expressed in a suitable background host, in which the activity of the enzyme was shown to catalyse the synthesis of indigo[27] (Figure 7).

Stabilisation of biocatalytic systems may be achieved by a variety of methods, all with a different degree of technical complexity: (i) stabilisation by use within the cellular environment, as described earlier; (ii) stabilisation by immobilisation of the system to a suitable support[28,29]; (iii) stabilisation through the application of protein engineering[30]: if the structural parameters which confer stability upon a protein, such as the intermolecular bonding, are understood then it may prove possible to engineer these parameters into a target protein without making deleterious alterations in its catalytic capabilities.

2.7 Alteration of reaction characteristics by alteration of reaction environment.

Discussion of biocatalytic reactions has centred so far upon reactions that occur in predominantly aqueous environments and that occur essentially in one direction only. Many biological reactions are reversible in nature and the direction in which they catalyse the reaction is dependent upon the relative concentrations of the reactants and the products. Reactions may be forced in one

Figure 6. Chemoenzymatic process for the production of polyphenylene[26]

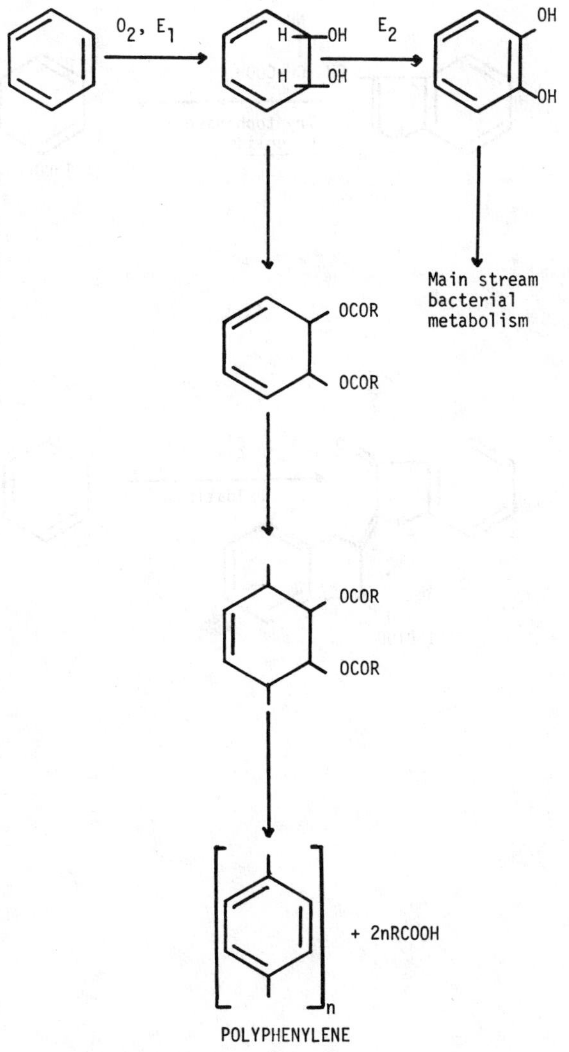

O_2, E_1

H OH

H OH

E_2

OH

OH

Main stream
bacterial
metabolism

OCOR

OCOR

OCOR

OCOR

+ 2nRCOOH

n

POLYPHENYLENE

Figure 7. Biosynthesis of indigo by a genetically engineered *E. coli*.[27]

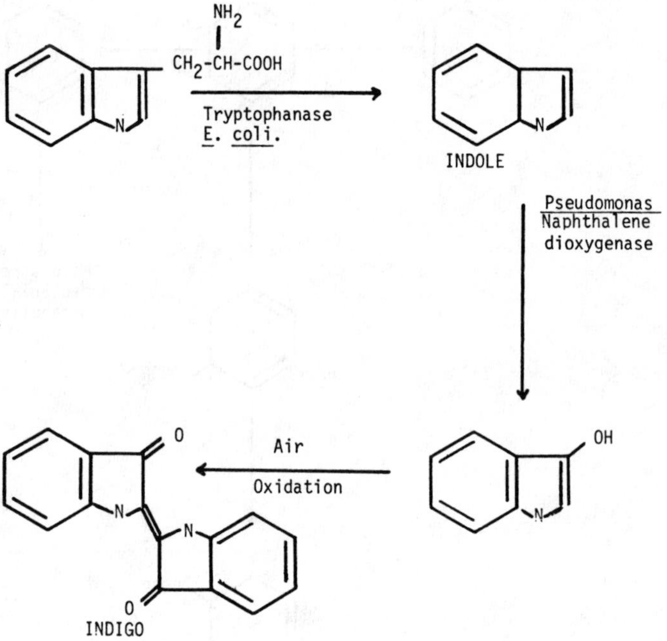

NH_2

CH_2-CH-COOH

Tryptophanase
E. coli.

INDOLE

Pseudomonas
Naphthalene
dioxygenase

OH

O

Air

Oxidation

N

N

N

O
INDIGO

direction by the removal of the product, ensuring that the equilibrium is in favour of its production and that the forward reaction does not slow down due to the stimulation of the reverse reaction. However, alteration of the reaction conditions has been successfully utilised to coerce an enzymatic reaction in the reverse direction to that in which it normally operates. Lipases are used in an aqueous environment to catalyse the hydrolysis of triglycerides[31]. The reaction proceeds in the direction of hydrolysis due to the overwhelming concentration of one of the reactants, water. However, these enzymes are stable and functional in environments with a very low water activity and, in such environments, will catalyse (inter)esterification in preference to hydrolysis[31,32] (Table 3). The use of biocatalytic systems in essentially non-aqueous environments, especially in systems in which the reactants have low water solubilities, is under intense investigation. Non-denaturing organic reaction systems can be then used to substantially increase substrate availability and may have a dramatic effect upon reaction kinetics. Examples in which such organic environments have been evaluated include cholesterol oxidation by Nocardia species[33] and the resolution of menthol esters[34] (Table 4). Proteins are thought to maintain their stability in these reaction systems by maintaining a discrete water layer around the protein molecule[35]. Micellar and emulsion enzyme technology[36] will see dramatic advances of fundamental importance to the chemical industry in the next ten years. Such an approach is a true example of innovative technology, as it applies an alternative logic to develop a new generation of biocatalysts in environments that might be thought of as bioincompatible.

2.8 Reaction configuration

An essential consideration in the implementation of biotechnology to chemical processes involves the design and scale up of the appropriate reactor configuration, which will enable the reaction to proceed in a controlled and predictable fashion. The application of chemical engineering to biotechnological processes is outside the scope of this contribution but it is pertinent to observe that chemical engineering technology has and will make an

TABLE 3: Esterification and interesterification reactions using lipases

(a) Esterification

Reactants:	Terpene alcohols and short and medium chain length fatty acids.
Catalyst:	Cell wall bound lipase from Rhizopus arrhizus.
Environment:	Substrates dissolved in diisopropyl ether or n-heptane.

(b) Interesterification

Reactants:	Mixtures of triglycerides or triglycerides and free fatty acids.
Catalyst:	Non-specific lipases, catalysing reactions resembling chemical interesterification. 1,3 specific lipases, catalysing interesterification at the 1 and 3 positions only, a process which is not possible chemically.
Environment:	Low water activity

: triglyceride

OH
OH : free acid

\+ → + +

\+ OH → + +

OH OH

TABLE 4: Biocatalysis in organic environments

(a) Asymmetric hydrolysis of menthol esters

Organisms used:	Absidia, Penicillium, Rhizopus, Trichoderma, Bacillus, Pseudomonas, and Rhodotorula species.
Esters used:	Acetates, monochloroacetates, lactates, propanoates, caproates and succinates.
Productivity:	A Rhodotorula mucalaginosa mutant was reported to produce 44.4 g of (-)-menthol in 24 hours from a 30% menthol acetate mixture.
Organic reaction conditions:	Polyurethane-entrapped yeast in a water saturated n-heptane environment catalysed a 72.6% conversion of (dl)-menthol succinate and the biocatalyst had a half life of approximately 60 hours.

(b) Conversion of cholesterol to cholest-4-ene-3-one

Organism used:	Nocardia species NCIB 10554
Environment:	Carbon tetrachloride
Productivity:	7 g per day

enormous contribution to the realisation of 'the biotechnological potential' and that biological processes should be designed with process configuration very much in mind. For instance, if the process is to be operated at thermophilic temperatures then the rationale should identify the biocatalyst that can perform the reaction under those conditions, rather than identifying the catalyst that can perform the desired reaction and subsequently examining whether it will be amenable to the desired reactor conditions.

There are a multitude of reactor designs available to the biotechnologist but these are usually variations on three main themes:

- surface fermentation: one of the oldest traditional biotechnological methods of chemical production by growth-associated biocatalysis, whereby the organism is allowed to grow on the solid substrate (e.g. the koji process for citric acid production)[3] and the solids are extracted to recover the desired product after the optimum time for growth and production has lapsed. A slight technical innovation on this process is the surface fermentation, where the organism grows on the surface of the fermentation liquor and the product is removed to the aqueous medium. Both these processes are technically awkward because it is difficult, if not impossible, to control the reaction conditions and thus dictate the precise course of the catalysis.
- towards the middle of this century techniques for submerged culture were developed for the generation of chemicals[3]. The reaction mixture, the fermentation broth, is agitated at sufficient rate to ensure equal distribution of the organism (catalyst) as it grows and also to ensure that concentration gradients of substrate, oxygen and other nutrients do not build up and lead to local variations in reaction conditions. Sampling and determination of physical and chemical parameters of the reaction becomes much more feasible and, thus, leads to more precise reaction control. However, precise control is still not possible with whole organism catalysis due to the compartmentalisation of the reaction sites. Well-agitated reaction mixtures more closely approach the ideal for process

control of free solution biocatalytic reactions but other considerations, such as catalyst reuse and recycle, enter the economic equation at this point. Whole organism catalysis (whether growth-associated or post-growth) can also be conducted under conditions of catalyst recycle (e.g. activated sludge digesters)[37] to increase catalyst concentration within the reactor and to reuse unspent catalytic potential. However, with increasing catalyst concentration, it becomes even more difficult to control precisely the local parameters under which the reactions are occurring and, in extreme circumstances, can lead to catalyst 'death', an extreme case of irreversible catalyst poisoning. The shear forces to which the biological entities are subject in large stirred batch reactors are also to be considered. Many of the genetically engineered catalysts may simply not be stable under the selection pressures imposed by the reactor conditions and therefore will not effectively perform in the biocatalytic process (e.g. the engineered 'Pruteen' organism).

- a significant advance in the technology and engineering of biological catalyst occurred with the advent of immobilised cell and enzyme technology. It opened up an exciting but already well characterised chemical engineering field of packed and fluidised bed reactors. This technology is only just beginning to be introduced into commercial biotechnology (the prime examples being waste treatment processes, glucose isomerase[12,13] and lactase[38]), as many biological systems have not been characterised with regard to the most suitable methods of immobilisation and the altered reaction kinetics produced as a result. However, the potential of this type of reactor configuration is enormous, using a reusable catalyst to generate a clean product stream onto which precise control may be imposed, once the characteristics of the system have been delineated.

3. Current applications of biotechnology to chemical production

The foregoing discussion has sought to highlight those aspects of biocatalysis that require consideration when applying biological systems to chemical production. This has been illustrated with several examples from current biological applications. Although biotechnology has made significant contributions to chemical production early this century, development was heavily overshadowed after the Second World War by the rapid expansion of the petrochemical industry. The technology was sustained and advanced in the food and pharmaceutical industries. However, the rapid developments in genetic and biochemical engineering witnessed over the past fifteen years have lead to a renewal of interest in the use of biological systems for novel chemical processes. The scale of the chemical process will affect the overall viability of the use of biocatalysis. The following section describes briefly the various scales on which current biological processes operate.

For bulk chemical generation, biological processes were used prior to the growth of the petrochemical industry for the production of solvents, such as acetone and butanol[39], replacing the traditional process of the destructive distillation of wood during the Great War. The single largest bulk chemical produced by a biological process is that of methane, by the anaerobic digestion of sewage and municipal waste, a process of great complexity[37]. Although biogas is not used extensively in the western world, it is a source of fuel in developing countries such as China. The petroleum crisis of the early seventies shifted the emphasis of fuel production from petroleum to other renewable sources. This has lead to massive research programmes in the States and in Brazil to develop the commercial production of ethanol ('gasohol') by fermentation of sugar crops[2] (Table 5).

On the medium scale, in terms of total production, are the traditional biocatalytic processes for the production of organic acids by fermentation[3] and the well-characterised and sophisticated industrial procedures for amino acid production, developed extensively in Japan[40]. Recent biotechnological processes added to this list include the production of single cell protein of bacterial (ICI Pruteen for animal feed)[41] and fungal (RHM MycoProtein for human consumption)[42] origin and polymer production[43] (both polysaccharides for a wide variety of food and lubrication applications and poly-β-hydroxybutyrate for

TABLE 5: Production of ethanol for 'power alcohol'

Organisms used:	*Saccharomyces cerevisiae*, *S. uvarum*, *Candida utilis*, *Schizosaccharomyces pombe*, *Kluyveromyces lactis*, *K. fragilis*, *Clostridium thermosaccharolyticum*, *Zymomonas mobilis*
Substrates utilised:	Sucrose (beet/cane sugar, molasses), starch (corn, cassava), cellulose (wood, bagasse, straw), lactose (whey)
Pathway:	Anaerobic catabolism in yeast
Product concentration:	From 50 - 140 g/l, depending on strain and process configuration used

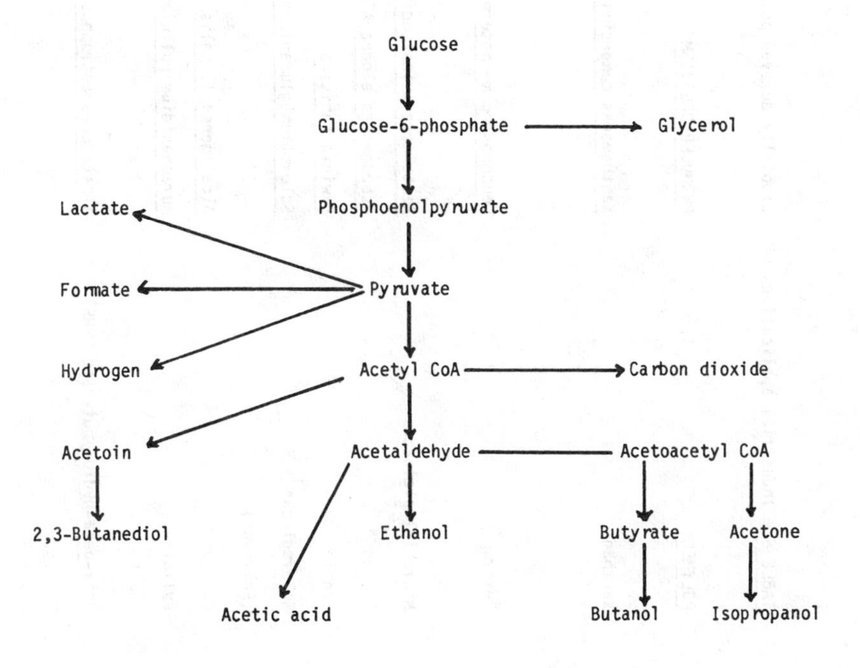

TABLE 6: Industrial applications of microbially derived polymers

POLYMER	PRODUCER ORGANISM	INDUSTRIAL APPLICATION
Xanthan	*Xanthamonas campestris*	Stabilisation and suspension agent, gelling and viscosity control in the oil and food industries
Dextran	*Leuconostoc mesenteroides*	Plasma substitutes and other medical applications
Microbial alginate	*Azotobacter vinelandii*	Thickening and gelling agent
Gellan	*Pseudomonas elodea* ATCC 31461	Agar and carrageenan replacement
Zanflo	*Erwinia tahitica*	Carpet printing
Scleroglucan (Polytran)	*Sclerotium glucanicum*	Drilling muds, latex paints, printing inks, seed coatings
Curdlan	*Alcaligenes faecalis* var. myxogenes	Gelling agent in food industry
Pullulan	*Aureobasidium pullulans*	Biodegradable films, fibres and packaging
Poly-*beta*-hydroxybutyrate (PHB)	*Alcaligenes eutrophus*	Biodegradable polymer: medical sutures

novel plastics with new properties: Table 6). The production of steroids and antibiotics is discussed separately in another chapter but can, of course, be included on this list. The application of many enzymatic processes to the starch processing and dairy industries may also be included in this category, although the individual catalytic steps are used to contribute to an overall more efficient biocatalytic fermentation process, to make better use of available carbohydrate sources as in the saccharification of starch[44].

Perhaps one of the most promising areas for the generation of chemical products by biotechnological means is the area of high value added commodities such as speciality polymers, novel sweeteners (such as aspartame), biologically derived flavours and fragrances and medical products that have been the target of so much of the recent genetic engineering technology (insulin, interferon, growth hormones etc).

An area which can be considered as chemical production involves the microbial reclamation of metals[45]; inorganic biocatalysis as opposed to organic chemical production (Table 7). The oxidative properties of certain classes of microorganisms may be used to assist the leaching of copper from low grade ore dumps economically and the ability of other microorganisms to selectively accumulate certain metals, such as uranium, is now under active investigation, for its potential in biologically mediated recovery processes. Biological strategies for the desulphurisation of oil and coal are also being evaluated[46]. In these circumstances biological processes may be the only economic route to chemical production and recovery.

4. Future applications

The future potential of biocatalysis in chemical production has been indicated but, in the long term, several infant technologies may come to the fore.

The derivation of chemicals from plant materials has sustained the chemical industry for many years, with the production of rubber, oils, alkaloids, fats and the like. However, these processes have always been subject to the

TABLE 7: Bioextractive metallurgy

Organisms used:	*Thiobacillus ferrooxidans*, *T. thiooxidans*, *T. acidophilus*, *T. organoparus*, *Sulfolobus*
Reactions catalysed:	$4FeSO_4 + O_2 + 2H_2O \longrightarrow 2Fe_2(SO_4)_3 + 2H_2O$ $S_8 + 12O_2 + 8H_2O \longrightarrow 8H_2SO_4$ $4FeS_2 + 15O_2 + 2H_2O \longrightarrow 2Fe_2(SO_4)_3 + 2H_2SO_4$ $ZnS + 2O_2 \longrightarrow ZnSO_4$ The ferric iron generated by this biological oxidation is then effective in oxidising other minerals: $Cu_2S + 2Fe_2(SO_4)_3 \longrightarrow 2CuSO_4 + 4FeSO_4 + S^0$ $UO_2 + Fe_2(SO_4)_3 \longrightarrow UO_2SO_4 + 2FeSO_4$
Other applications:	Metal accumulation by surface adsorption or intracellular accumulation, metal precipitation through the generation of organometallic complexes or the biological generation of hydrogen sulphide, metal volatilisation (e.g. methyl mercury)

vagaries of climate, local politics and variable product composition. Plant cell biotechnology may, over the next few decades, release the industry from its dependence on such a fluctuating market by two mechanisms: (i) improved production and quality control, using traditional methods of cropping, by the engineering of specific plant strains possessing the optimum physiological and biochemical characteristics for the production (e.g. cloning of oil palms[47]): (ii) through the development of plant cell technology, using free solution and immobilised reactor systems with particular plant cell lines to produce specific chemical products.

Alternative methods of synthesis of many natural secondary products are often not possible due the their inherent complexity and the market demand for naturally derived products. Expression of plant pathways in bacteria by genetic engineering is a long term goal that awaits a much more comprehensive understanding of the complex metabolic relationships that exist within these secondary metabolic pathways. The primary interest in this technology at present is for the generation of pharmaceutical products[48] (Table 8) and for the improvement of field crops.

A prime target of the organic chemist for many years has been the controlled site-specific introduction of oxygen into organic molecules. This has been very succesfully commercialised for the production of steroids but other industrial oxyfunctionalisation reactions have not been approached down the biotechnological route, primarily due to the instability of the oxygenase enzymes involved in the oxygen insertion reactions. However, there is currently much research effort being devoted to the use of microbes and enzymes for these conversions (e.g. epoxidation of propene[49]; Figure 8).

Another area of intense activity at present is that of protein engineering, whereby proteins are modified either by altering the genes coding for these proteins or by chemical modification of the protein itself, to effect changes in the stability or kinetics of the system. Protein engineering should eventually allow the molecular biologist and biochemist to tailor the catalyst to a particular set of requirements, within certain defined limits of biological

TABLE 8: Potential chemical products using plant cell biotechnology

(a) de novo synthesis

PRODUCT	PLANT CELL LINE
Ajmalicine	Catharantheus roseus
Serpentine	Catharantheus roseus
Vinblastine	Catharantheus roseus
Vincristine	Catharantheus roseus
Alkaloids	Nicotinia species
Capscicin	Capsicum
Shikonin	Lithospermum erythrorhizon

(b) Biotransformation

PRODUCT	PLANT CELL LINE
Digitoxin digoxin	Digitalis lanata (foxglove)

FIGURE 8: The Cetus process for the biocatalytic production of alkene epoxides

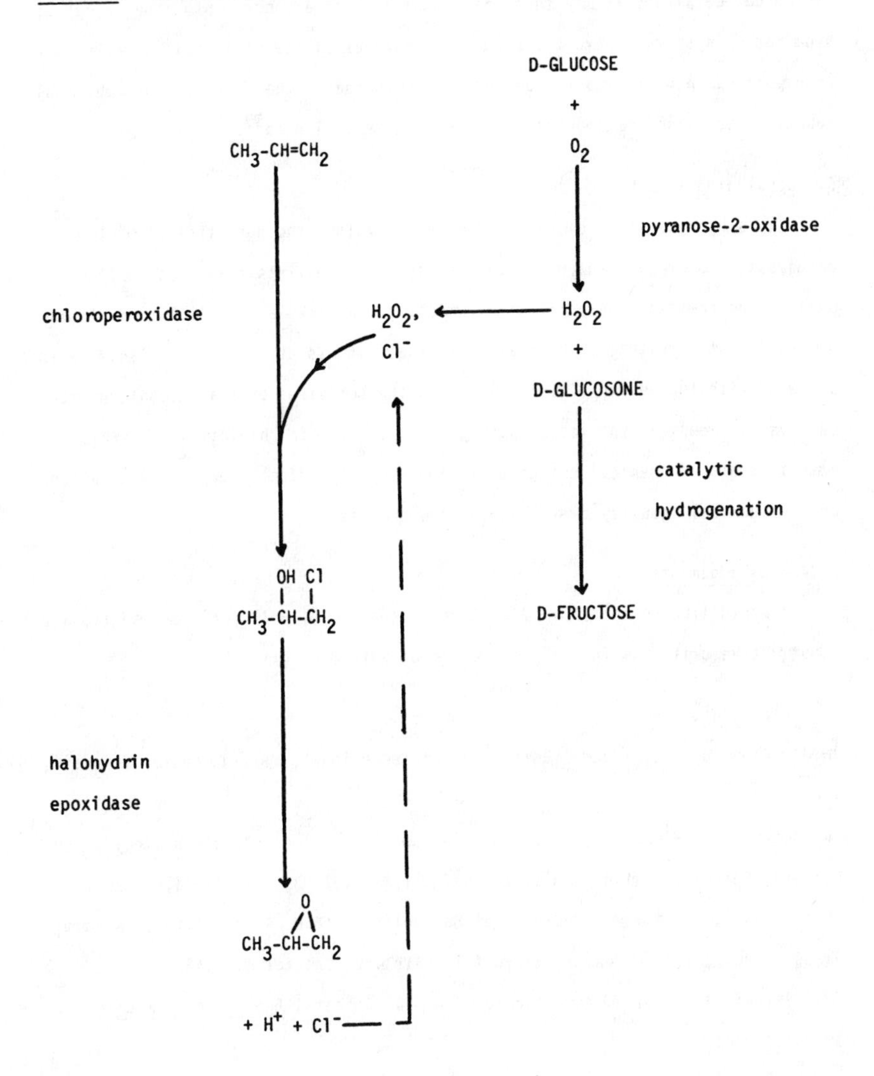

catalysis.

As a result of understanding the properties and mode of action of biological catalysts it may then be possible to advance into the area of biomimetic systems, where the essential features of the biological catalyst are incorporated into bioorganic complexes which possess the potency of chemical reactions but also the sophistication of biochemical ones[50].

5. Concluding remarks

It is necessary, in order to advance and fulfil the potential of biological catalysis, to regard enzymological catalysis, in all its forms, as an integral part of the chemist's weaponry with which to attack a particular synthetic process. At present, the state of the art is still in its infancy, when compared with the wealth of information on the behaviour of traditional chemical catalysts. However, the recent upsurge in interest in biological systems as potent agents of chemical change will significantly affect the shape and outlook of the chemical industry over the next few decades.

6. Acknowledgements

I would like to thank Dr. Alex Cornish and Dr. Al Hall for their advice and constructive criticism in the preparation of this manuscript.

7. References.

[1]H.J. Rehm and G. Reed (Editors), "Biotechnology", Verlag Chemie, Weinheim, 1983, Vol. 5.

[2]N. Kosaric, A. Wieczorek, G.P. Cosentino and R.J. Magee, "Biotechnology", (H. Dellweg, Editor), Verlag Chemie, Weinheim, 1983, Vol. 3, Chapter 3a, p. 257.

[3]L.M. Miall, "Primary Products of Metabolism", (A.H. Rose, Editor), Academic Press, London, 1977, Economic Microbiology Vol. 2, Chapter 3, p.48.

[4]M. Kitsumi, J. Kato, S. Komatsubara and I. Chibata, Appl. Env. Microbiol., 1970, 21, 569.

[5]S. Udaka and S. Kinoshita, J. Gen. Appl. Microbiol., 1958, 13, 303.

[6]M.A. Schell, J. Bact., 1983, 153, 822.

[7]A.M. Klibanov, Basic Life Sci., 1983, 25, 497.

[8]Y. Yamaguchi, A. Komatsu and T. Moroe, J. Agric. Chem. Soc. (Japan), 1977, 50, 443.

[9]P. Newman, "Optical Resolution Procedures for Chemical Compounds", Optical Resolution Information Centre, Manhattan College, New York, 1981, Vol. 2.

[10]K. Kieslich and O. Sebek, Ann. Rep. Ferm. Proc., 1979, 3, 267.

[11]N. Esaki, H. Tanaka, S. Uemura, T. Suzuki and K. Soda, Biochemistry, 1979, 18, 407.

[12]W.P. Chen, Proc. Biochem., 1980, 15, 30.

[13]W.P. Chen, Proc. Biochem., 1980, 15, 36.

[14]D.J. Best and I.J. Higgins, "Topics in Enzyme and Fermentation Technology", (A. Wiseman, Editor), Ellis Horwood Ltd., Chichester, 1983, Vol. 7, Chapter 3, p. 38.

[15]P.G. Wislocki, G.T. Miwa and A.Y.H. Lu, "Enzymatic Basis of Detoxification", (W.K. Jakoby, Editor), Academic Press, New York, 1980, Vol. 1.

[16]C. Anthony, "The Biochemistry of Methylotrophs", Academic Press, London, 1982.

[17]B. Atkinson and F. Mavituna, "Biochemical Engineering and Biotechnology Handbook", Macmillan Publishers Ltd., Byfleet, 1983.

[18]R. Uchio and I. Shiio, Agr. Biol. Chem., 1972, 36, 1389.

[19]W. Crueger and A. Crueger, "Biotechnology: A Textbook of Industrial Microbiology", Science Tech Inc., Madison, 1984, Chapter 11, p. 163.

[20]M.L. Hemming, J.C. Ousby, D.R. Plowright and J. Walker, Water Pollution Control, 1977, 76, 451.

[21]P.M. Sutton, W.K. Shieh, P. Kos and P.R. Dunning, "Biological Fluidised-Bed Treatment of Waste Water", (P.F. Cooper and B. Alkinson, Editors), Ellis Horwood, Chichester, 1981, p. 285

[22]J.E. Bailey and D.F. Ollis, "Biochemical Engineering Fundamentals", McGraw-Hill Book Company, New York, 1977, p. 84.

[23]S.K. Tangnu and T.K. Ghose, Proc. Biochem., 1980, 15, 30.

[24]H. Tone, A. Kitai and A. Ozaki, Biotech. Bioeng., 1968, 10, 689.

[25]J.D. Windass, M.J. Worsey, E.M. Pioli, D. Pioli, P.T. Barth, K.T. Atherton,

E.C. Dart, D. Byrom, K. Powell and P.J. Senior, Nature, 1980, 287, 396.

[26]D.G.H. Ballard, A. Courtis, I.M. Shirley and S.C. Taylor, J. Chem. Soc. Chem. Comm., 1983, 17, 954.

[27]B.D. Ensley, B.J. Ratzkin, T.D. Osshund, M.J. Simon, L.P. Wackett and D.T. Gibson, Science, 1983, 222, 167.

[28]O.R. Zaborsky, "Immobilized Enzymes", CRC Press Inc., 1973, Cleveland, Ohio.

[29]P. Monsan, "Biotech '84, Europe", Online Publications, Pinner, 1984, p. 379.

[30]K.M. Ulmer, Science, 1983, 219, 666.

[31]A. Kilara, Proc. Biochem., 1985, 20, 35.

[32]A.R. Macrae, J. Am. Oil Chem. Soc., 1983, 60, 291.

[33]B.C. Buckland, P. Dunnill and M.D. Lilly, Biotech. Bioeng., 1975, 17, 815.

[34]Ube Industries Ltd., Chem. Abs., 1981, 94, p. 14031.

[35]S.L. Neidleman, "Biotechnology and Genetic Engineering Reviews", (G.E. Russell, Editor), Intercept, Newcastle-on-Tyne, 1984, Vol. 1, Chapter 1, p. 1.

[36]C.A. Bunton, "The Chemistry of Enzyme Action", (M.I. Page, Editor), Elsevier, Amsterdam, 1984, Chapter 13, p. 461.

[37]A.D. Wheatley, "Biotechnology and Genetic Engineering Reviews", (G.E. Russell, Editor), Intercept, Newcastle-on-Tyne, 1984, Vol. 1, Chapter 10, p. 261.

[38]V. Gekas and M. Lopez-Leiva, Proc. Bioch., 1985, 20, 2.

[39]M.T. Walton and J.L. Martin, "Microbial Technology: Volume 1, Microbial Processes", (H.J.Peppler and D. Perlman, Editors), Academic Press, New York, 1979.

[40]K. Soda, H. Tanaka and N. Esaki, "Biotechnology Volume 3: Biomass, Microorganisms, Products I, Energy", (H. Dellweg, Editor), Verlag Chemie, Weinheim, 1983, Chapter 3g, p. 479.

[41]S.R.L. Smith, Phil. Trans. R. Soc. Lond., 1980, B290, 341.

[42]G.A. Beech, M.A. Melvin and J. Taggart, "Biotechnology, Principles and Applications", (I.J. Higgins, D.J. Best and J. Jones, Editors), Blackwell Scientific Publications Ltd., Oxford, 1985, Chapter 3, p.73.

[43]R.C.W. Berkeley, G.W. Gooday and D.C. Ellwood (Editors), "Microbial Polysaccharides and Polysaccharases", Academic Press, London, 1979.

[44]T. Harada, "Biotechnology and Genetic Engineering Reviews", (G.E. Russell, Editor), Intercept, Newcastle-on-Tyne, 1984, Vol. 1, Chapter 2, p. 39.

[45]L.E. Murr, A.E. Torma and J.E. Brierley (Editors), "Metallurgical Applications of Bacterial Leaching and Related Microbiological Phenomena", Academic Press, New York, 1978.

[46]W.R. Finnerty and F.J. Hartdegen, "Biotech '84, Europe", Online Publications, Pinner, 1984, p. 611.

[47]L.H. Jones, Biologist, 1983, 30, 181.

[48]E.J. Staba (Editor), "Plant Tissue Culture as a Source of Biochemicals", CRC Press, Boca Raton, Florida, 1980.

[49]S.L. Neidleman and J. Geigert, "Biotechnology", (C.F. Phelps and P.H. Clarke, Editors), Biochemical Society Symposium No. 48, Biochemical Society, London, 1983, p. 39.

[50]T.H. Maugh, Science, 1984, 223, 154.

12

Enzyme Recovery and Purification: Downstream Processing

By T. Atkinson, P. M. Hammond, M. D. Scawen, and R. F. Sherwood

MICROBIAL TECHNOLOGY LABORATORY, P.H.L.S. CENTRE FOR APPLIED MICROBIOLOGY AND RESEARCH, PORTON DOWN, SALISBURY, WILTS. SP4 0JG, U.K.

INTRODUCTION

Enzymes are increasingly employed as reagents in clinical chemistry, as therapeutic agents in chemotherapy and in industrial processes. The much publicised advances in molecular genetics have resulted in an awareness of the importance of protein recovery and purification. Downstream processing is therefore best described as the recovery and purification of the desired protein product from hundreds or thousands of litres of microbial culture.

It is necessary to define large scale: the great majority of enzymes in industrial use are extracellular proteins from Aspergillus sp. or Bacillus sp. and include α-amylase, β-glucanase, cellulases, dextranase, proteases, glucoamylase, etc.[1] For such enzymes, the degree of purification required is minimal and the normal scale of operation yields tonnes of protein product. Many other enzymes are required for non-industrial uses. These are intracellular and are produced in much smaller amounts, for example asparaginase, catalase, cholesterol oxidase, β-galactosidase, glucose oxidase and glucose-6-phosphate dehydrogenase. These enzymes are typically produced in kilogram quantities from hundreds or a few thousand litres of culture. They are also generally purified to homogeneity; hence purification procedures are more demanding and final recovery lower. In this article we concentrate on techniques suitable for the isolation and purification of bacterial enzymes from 1 - 200 kg starting material. This scale is sufficient for trial quantities of a new enzyme and can in many cases involving rDNA products satisfy the total market demand.

Apart from scale, the major difference between isolating an intracellular and an extracellular enzyme is that cell lysis is required. Once the cells have been lysed, so as to release the intracellular protein, the same types of purification technique are applicable. Figure 1 illustrates the major steps involved in isolating and purifying an intracellular enzyme, from any source. The basic differences between bacterial, fungal, plant or mammalian sources relates to the ease of cell breakage.

CELL LYSIS

Chemical Methods of Cell Lysis

Alkali. This method has been used with considerable success in small and large-scale extractions of bacteria. Wade[2] isolated the therapeutic enzyme L-asparaginase by exposing *Erwinia* sp. to an alkaline pH between 11.0 and 12.5 for 20 minutes. The success of alkali treatment is dependent on the stability of the required enzyme to high pH. It has been suggested that this method may help inactivate proteases and reduce the possibility of pyrogen contamination of therapeutic enzyme preparations. The method is also applicable to rapid inactivation and lysis of rDNA microorganisms.

Lysozyme. Lysozyme, an enzyme produced commercially from hen egg white, specifically catalyses the hydrolysis of β-1-4-glycosidic bonds in the mucopeptide of bacterial cell walls. Gram-positive bacteria, which rely on wall mucopeptide for rigidity, are most susceptible to lysozyme. However, final rupture of the cell envelope often depends upon the osmotic pressure of the suspending medium once the cell wall has been digested. In Gram-negative bacteria the breakdown of the cell wall is rarely achieved by lysozyme alone, but the addition of EDTA to chelate metal ions will frequently result in lysis. This technique is rarely used for the large scale extraction of bacterial enzymes due to the relatively high cost of lysozyme although practically this is small in comparison with the value of the purified enzyme. Crude hen egg white is much cheaper and often equally effective. Experiments have been carried out with immobilised lysozyme, but with limited success.[3]

Figure 1

Stages in the isolation of an intracellular enzyme

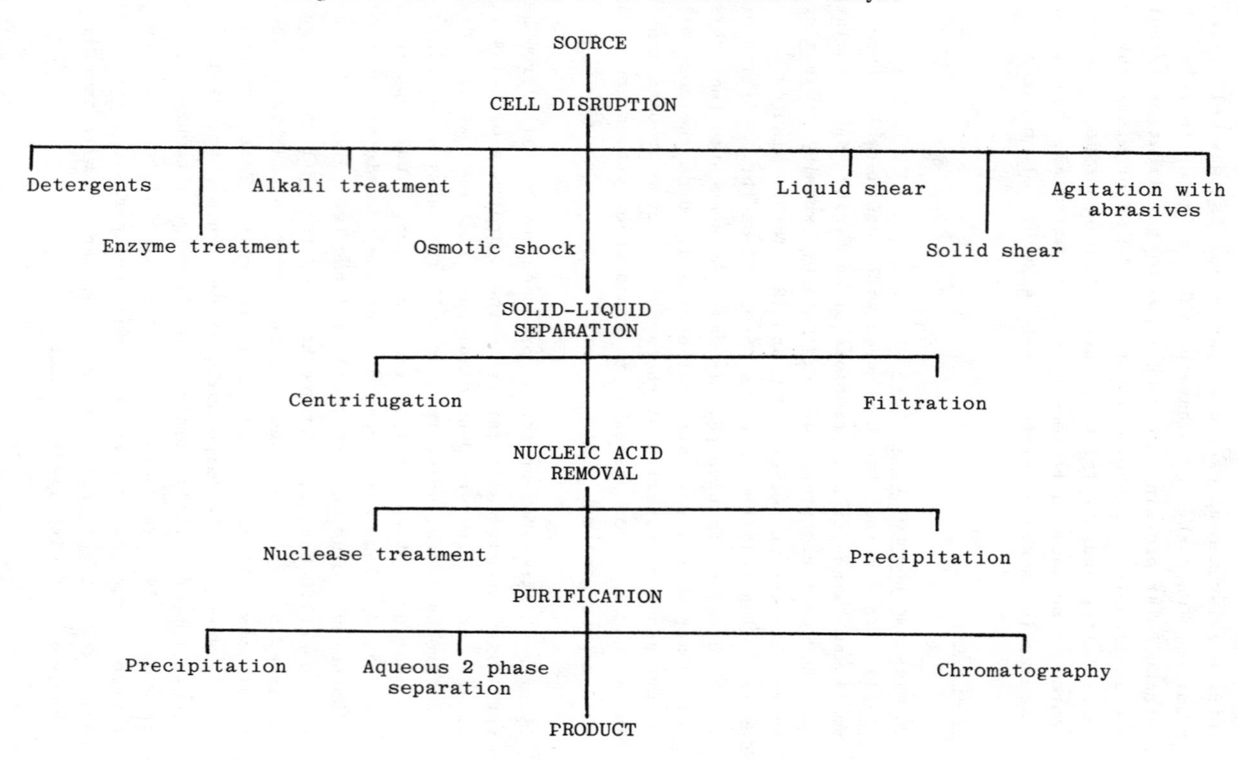

Detergents. Detergents, either ionic, for example sodium lauryl sulphate, sodium cholate (anionic) and cetyldiethylammonium bromide (cationic), or nonionic, for example the Tritons, have been used to aid cell lysis. Ionic detergents are more reactive than non-ionic detergents and can lead to the dissociation of lipoproteins, which can further lead to protein denaturation; for this reason such detergents are not desirable in the extraction of enzymes. The presence of detergent can also affect subsequent purification steps, in particular salt precipitation of proteins. This can be overcome in many cases by the use of ion exchange chromatography or ultrafiltration but obviously necessitates additional steps. Nevertheless, detergents do have considerable use in some extraction processes. Triton X-100 has been used for the large scale release of cholesterol oxidase from *Nocardia* sp.[4] and sodium cholate used to solubilise pullulanase (pullulan-6-glucan hydrolase), a membrane-bound enzyme from intact cells of *Klebsiella pneumoniae*[5].

Physical Methods of Cell Lysis

Sonication. The term ultrasonics is used to denote frequencies of 20kHz and above. The application of ultrasonics in liquid creates cavitation. Areas of compression and rarefaction occur, and cavities, which form in the areas of rarefaction, rapidly collapse as the area changes to one of compression. The bubbles produced in the cavities are thus compressed to several thousand atmospheres. On their collapse shock waves are formed, and these shock waves are thought to be responsible for the damage to the cells. There is considerable variation in the susceptibility of microorganisms to sonication. In general, Gram-negative organisms are more susceptible than Gram-positive organisms and rod-shaped bacteria are more readily broken than cocci. In addition, the efficiency of ultrasonics depends upon environmental parameters including pH, temperature, ionic strength of the suspending medium and the time of exposure and cell density. The selection of conditions is essentially empirical and varies with the particular organism as well as the required product. Although ultrasonic treatment has proved to be a useful and versatile method for laboratory-scale work, its application to the disruption of large quantities of bacteria is limited by the difficulties of transmitting sufficient power to large volumes of suspension and of removing the heat generated.

Osmotic shock. Osmotic shock has been used in the extraction of hydrolytic enzymes and binding proteins from the periplasmic space of a number of Gram-negative bacteria, including *Salmonella typhimurium* and *E. coli*[6]. The method involves washing the bacteria in buffer solution to free them from growth medium and then resuspending them in, for example, 20% buffered sucrose. The cells are then allowed to equilibrate, resulting in the loss of some internal water and are then removed from suspension. The cell paste obtained is then rapidly dispersed in water at approximately 4°C.

Only 4-8% of the total bacterial protein is released by osmotic shock. If the required enzyme is located in the periplasmic region osmotic shock can produce a 14- to 20-fold increase in purification compared to other disruptive techniques[7]. Osmotic shock has proved to be particularly useful for the extraction of periplasmic aminoglycoside inactivating enzymes, *e.g.* kanamicin acetyl transferase from *E. coli*[8], and is also a valuable technique for the release of enzymes from marine bacteria such as luciferase from *Photobacterium fischeri*[9] since it is only necessary to resuspend cells of these organisms in dilute buffer to achieve lysis as the growth medium contains 0.5M NaCl.

Freezing and thawing. The effects of freezing and thawing on microorganisms are similar to those observed during osmotic shock. The efficiency for general protein release is however limited and less than 10% of the total soluble protein is normally liberated, even from Gram-negative bacteria. Proteins located in the periplasmic space can be released more readily, however, and up to 60% of a R-factor mediated penicillinase from *E. coli* appears in the supernatant after the cells had been frozen and thawed once (Melling, pers. comm.). Although the technique has the advantages of simplicity and low temperature operation, it is not extensively used because it is time consuming. Many organisms are resistant to rupture and many enzymes are inactivated by repeated freezing and thawing. Freezing and thawing is often, probably unintentionally, an adjunct to other cell-breaking methods since bacterial pastes are often stored at $-20^{o}C$ prior to enzyme extraction and purification. The X-press is capable of processing 100g of Saccharomyces cerevisiae per minute[10]. Frozen cells are disrupted by passage through a perforated disc, the outlet temperature being approximately $-22^{o}C$. Rupture of the cells is caused by the shear forces exerted by the passage of the extruded paste through the small orifice, the shear being aided by ice crystal formation in the frozen paste. A larger version of the solid-shear disintegrator has been developed by Magnusson and Edebo[11], which is capable of disrupting up to 10kg yeast paste per hour.

Grinding with abrasives. Initially this technique was restricted to the grinding of cell pastes in a mortar with an abrasive such as glass, aluminium or kieselguhr. This system has since been developed and mechanised using glass beads - the Mickle or Braun shakers. Such devices were inefficient and restricted to small amounts of material. Modern, larger scale versions are based on machines developed for the wet grinding and dispersion of pigments in the paint and allied industries and provide an efficient means of disintegrating large amounts of bacteria.

A typical commercially produced apparatus, the Dynomill, (W.A. Bachofen, Switzerland) has produced a most efficient method for the disruption of some 'tough' bacteria including Streptococcus mutans and Staphylococcus aureus[12]. Up to 5kg per hour of such bacterial paste can be treated using the laboratory scale machine with a 600ml continuous-flow grinding container. Marfy and Kula[13] have studied the release of enzymes from brewer's yeast under various conditions. Rehacek and Schafer[14] studied the disintegration of yeast cells in a novel type of agitator in which the agitator discs were placed alternatively perpendicularly and obliquely on the drive-shaft, an arrangement claimed to be more efficient. Recently, Woodrow and Quirk[15] examined conditions necessary for the optimum release of a number of bacterial enzymes using a Dyno-Mill model KPL fitted with a 600ml continuous flow chamber. They found that this small unit could handle up to 4kg bacteria per hour. Extrapolation from this observation would indicate that the larger models could handle up to 100kg bacteria per hour.

In general, the rate of cell breakage is dependent on a variety of factors: the size and concentration of the beads; the type, concentration and age of the cells; the agitator speed; the flow rate through the chamber; the temperature of disruption; and the arrangement of the agitator discs. The Dyno-Mill type of cell disruptor has the significant advantage that it can be readily mounted in a safety cabinet when pathogenic or rDNA micro-organisms are to be disrupted.

Liquid shear. This is now the principal mechanical method of cell breakage used in the large scale disruption of micro-organisms[16]. The basis of the technique is that a suspension of bacteria is forced through a small orifice under very high pressure. As is commonly the case in enzyme processing, the equipment employed was originally designed for other purposes; the high pressure homogenisers are typified by the APV Manton-Gaulin homogeniser being developed for the production of emulsions in the dairy industry.

The APV Manton-Gaulin homogeniser is a positive-displacement pump, incorporating an adjustable valve with a restricted orifice. The cells in liquid suspension are passed through the homogeniser at pressures up to 55MPa. Charm and Matteo[7] indicated the three ways in which the homogenisers can be used; single pass, batch recycle, and continuous recycle and bleed. They also gave equations for determining the number of whole cells remaining and the overall processing rate of the method. Protein release on disruption of yeast cells is described by the first order rate equation[17]

$$\text{Log}\ (Rm/Rm - R) = KNP^{2.9}$$

where

R = amount of soluble protein released in g per g of packed cells

Rm = the maximum amount of soluble protein released

K = a temperature-dependent rate constant

N = the number of times the suspension is passed through the machine

P = operating pressure

2.9 = power variation

Protein release was found to be coincident with cell disruption, although some enzymes were released more slowly and others more rapidly than the mean protein release rate. These differences were not sufficient to achieve significant purification and, although rate equations may be used to describe protein release under well-defined conditions, in practice their value is limited.

Gram-negative bacteria are, in general, more easily disrupted than Gram-positive organisms[7] and the history of the material, including such factors as growth conditions[18] and freezing and thawing, affects the protein release rate. In particular, periplasmic enzymes are readily liberated[19].

The liquid shear homogeniser is normally operated at wet cell concentrations around 20% and the smallest APV Manton-Gaulin homogeniser, the 15M-8BA, has a throughput of 54 litre hr^{-1} at 55 MPa. A larger version, the K3 is also available, with a throughput of 250 litre hr^{-1} at 35MPa. This lower operating pressure means that the cell suspension must be recycled to ensure optimum breakage and cooling is required as temperature rises of 15-20° occur on passage through the disruption valve.

Nucleic acid removal. Nucleic acids can be removed by precipitation with a variety of positively charged compounds such as cetyltrimethylammonium bromide, streptomycin sulphate, protamine sulphate, polyethyleneimine, polylysine or $MnCl_2$. However, in recent years, the preferred method has been nuclease treatment[20]. The addition of two concentrations of crude pancreatic deoxyribonuclease to the crude extract causes a rapid reduction in viscosity and increases the efficiency of subsequent centrifugation or filtration steps. In addition to its rapidity, nuclease treatment is very significantly cheaper than all the other techniques and does not carry any risks such as co-precipitation of proteins, that can occur when nucleic acids are precipitated. It cannot be used however when DNA utilising enzymes are the objective of the purification!

ISOLATION AND INITIAL PURIFICATION

Debris removal. Following cell breakage and nucleic acid removal, the first step in the purification of an intracellular enzyme is the removal of cell debris. The separation of solids from liquids is a key operation in enzyme isolation, and is normally accomplished by centrifugation or filtration. A variety of centrifuges are available for large scale enzyme purification:

Batch centrifuges. Batch centrifuges are available with capacities ranging from less than 1 ml up to several litres and capable of applying a centrifugal force of up to 100,000 x g. However, for the deposition of bacterial cells, cell debris or protein precipitates forces up to 20,000 x g are usually sufficient. Machines capable of this and of handling large quantities of material are less numerous.

Continuous flow centrifuges. In the early stages of large-scale enzyme purification processing it is necessary to remove solids from several hundred litres of suspension. In this type of centrifuge, the deposited solids are retained in the centrifuge bowl and the clarified supernatant is continuously discharged. There are three basic designs of centrifuge which can retain several kilograms of solid and operate at sufficiently high flow rates: the disc type centrifuge, the tubular bowl centrifuge and the basket centrifuge.

The bowl used with the former machines contains a central stack of coned discs ensuring an almost constant length of flowpath with the deposition of solids at the bowl wall producing only a minor reduction in the flowpath. Thus, there is little loss of efficiency as the process proceeds. Cleaning is, however, laborious and loss of deposit is difficult to avoid and re-assembly is time consuming.

A variation of the disc type centrifuge is the multi-chambered bowl centrifuge, in which the bowl is divided by vertically mounted cylinders into a number of interconnected chambers. The feed passes through each chamber from the centre outwards before leaving the centrifuge. This type of arrangement ensures a short and constant settling path as the bowl fills and is easier to dismantle and clean than the disc-bowl type. In some centrifuges, the deposited solids can be discharged intermittently as a slurry without stopping the machine. However, experience with centrifuges of this type has indicated that some losses of enzyme activity occur when the solid is discharged. Nevertheless, disc type centrifuges with solid discharge have proved efficient for clarification and the use of this type of discharge can significantly reduce process labour costs.

Centrifuges of these types are available from De Laval Separator Co., New York: Westfalia Separator Ltd., Wolverton, Bucks; Bird Machine Co., South Walpole, Massachusetts and have solid capacities of up to 60 kilograms operating at about 8000 x g, although newer machines are available which generate 14,000 x g. The selection of a suitable flow rate is essentially empirical and varies greatly with the nature of the feed material.

Hollow bowl centrifuges. Such have a tubular bowl in order to obtain a sufficient length of flow path. Thus as centrifugation proceeds the effective bowl diameter, and hence the centrifugal force, are reduced. However, such centrifuges are easily cleaned and by using a bowl liner the recovery of deposited material is virtually 100%; re-assembly is also a quick and simple procedure. The Pennwalt centrifuges are perhaps the best known of this type (Table 1) although similar machines are also available from Carl Padberg, Lahr, West Germany. Flow rates must be determined empirically, but for deposition of cell debris after bacterial disruption flow rates of about 50 litre/hr have been used with the larger machines.

Table 1

Characteristics of Pennwalt Tubular Bowl Clarifiers

Model	Bowl capacity (g wet deposit)	Maximum RCF	Bowl weight
T-41-24Y	200	50,000 or 13,000	1.4kg
A.S.16	3,500	13,000	27.0kg
A.S.26	5,200	16,000	63.0kg

Basket centrifuges. These are designed to operate at much lower g forces than the centrifuges described above (1,000 rpm compared with 10,000 rpm) and are in essence centrifugal filters, the bowl being perforated to allow egress of filtrate. Porous bowl liners of some filter cloth are normally used, the main purpose of these machines being to collect large particulate material. In the context of enzyme purification this usually means ion exchange cellulose or resins. Basket centrifuges may be useful for the collection of materials during batch absorption of enzymes; they are also useful when ion exchangers are being regenerated or equilibrated but have also been used for the assisted removal of cell debris with filter aids such as lightly coated DEAE-cellulose.

Perhaps the simplest example is the ordinary domestic spin dryer. Its main drawbacks are low capacity and the inability to operate on continuous flow. Basket centrifuges of various sizes are available from Carl Padberg, Lahr, West Germany and Thomas Broadbent, Huddersfield, U.K. which will operate continuously and can contain several tens of kilograms of ion exchangers.

Tangential Flow Filtration

Although centrifugation is the widely accepted method for removal of cell debris (and protein precipitates) it is not ideal for large scale use. Thus other techniques have been researched, and filtration, although used in many industries for solid-liquid separations, is poorly suited to the clarification of bacterial (or animal homogenates) and the removal of protein precipitates, which are often gelatinous in nature and filter poorly unless the precipitation conditions can be adjusted to give a flocculent precipitate.

An alternative method of preventing the blocking of filtration membranes is tangential flow, in which the liquid flow is at right angles to the direction of filtration. Thus at a sufficiently high flow rate blockage of the filter can be minimised. Quirk and Woodrow[21] demonstrated with a Millipore Pellicon cassette system the potential of tangential flow filtration for clarification of two bacterial extracts;

Pseudomonas fluorescens containing aryl acyl amidase[22] and a Pseudomonas sp. containing carboxypeptidase G_2[23]. These results showed that both the organism and the method of cell breakage had a marked effect on the filtration rate. In addition, the isotropic membranes used were found to be prone to blocking by proteins and debris. The recovery of enzyme could be improved by diluting the extract and by washing the membranes. However, these steps gave a considerable increase in volume.

A new type of membrane, with an asymmetric structure, has recently been made available by Domnick Hunter Filters Ltd. (Birtley, Co. Durham, U.K.). These membranes are much less prone to fouling and hence can handle much higher concentrations of solids. Using the aryl acylamidase from Pseudomonas fluorescens the effects of pressure, feed velocity and membrane configuration were investigated with a unit containing a total of $128cm^2$ of membrane[24]. These same membranes were also tested on a larger scale employing Erwinia chrysanthemi. A 1 m^2 membrane assembly was used to harvest the cells from 100 litres of culture fluid in 2.5h. The solids concentration in the retentate increased from 0.55% dry weight to 22% dry weight. This same membrane assembly was then used to clarify the extract produced by the alkali lysis of these bacteria[24]. The date indicated that to harvest the bacteria from a 500 litre culture in 2.5h would require a total of $7.5m^2$ membrane. The economics of the membrane filtration process thus compare very favourably with centrifugation. In addition, the problems of aerosol generation are effectively eliminated, which is an important consideration when pathogenic or rDNA organisms are employed.

PRECIPITATION

Ammonium sulphate. Salting out of proteins has been employed for many years and fulfils the dual purposes of purification and concentration. The most commonly used salt is ammonium sulphate, because of its high solubility, lack of toxicity to most enzymes, low cost and, in some cases, its stabilising effect on enzymes[25].

The precipitation of a protein by salt depends on a number of factors: pH, temperature, protein concentration and the salt used[7]; particularly important in scaling-up protein purification is protein concentration. Large-scale enzyme purification is normally carried out at much higher protein concentrations than laboratory scale purifications, due to the need to maintain reasonable volumes. This can have dramatic effects on the concentration of a salt required to precipitate a given protein. The use of ammonium sulphate in enzyme purification has been described by many authors, the purification of penicillinase from E. coli strain W3310[19] being a typical example.

Organic solvents. The addition of organic solvents to aqueous solutions of proteins reduces the solubility of the proteins by reducing the dielectric constant of the medium. Temperatures above 4°C can lead to protein denaturation in the presence of organic solvents and thus it is desirable to work at temperatures below 0°C, which is feasible due to the depression of the freezing point produced by organic solvents. Various organic solvents have been employed for the precipitation of proteins, with ethanol and propan-2-ol being the most important, although acetone and even diethyl ether can be employed. These latter two solvents have the serious disadvantage of greatly increased flammability. For large-scale work, the precipitation and concentration of enzymes with organic solvents has not been routinely used. The flammable nature of the materials together with their relatively high cost has made them less attractive than other methods, with the notable exception of serum fractionation. In this process, despite the introduction of other techniques, ethanol precipitation has reigned supreme and the process has been developed into a highly automated computer-controlled system[26].

High molecular weight polymers. Other organic precipitants which can be used for the fractionation of proteins are the water soluble polymers of which polyethylene glycol is the most widely used. This has the advantages of being non-toxic, non-flammable and non-denaturing to proteins. Its mode of action and applications have been described[27,28] but it is not yet widely used, except in the blood processing field.

CHROMATOGRAPHY

The purification of enzymes by chromatography has been a laboratory practice for many years. These same chromatographic procedures can equally be employed for the isolation and purification of much larger quantities of enzyme. Thus, gel filtration, ion exchange, hydrophobic interaction and affinity chromatography have all been applied to the large-scale purification of a wide range of enzymes (Table 2). For large-scale work it has proved important to consider the order in which the steps are applied. On the laboratory scale, it has been a common practice to perform an initial fractionation of crude extract by gel filtration and follow this with one or more steps of ion exchange chromatography; with tens or hundreds of litres of extract this approach is impracticable. It is often better to start with an ion exchange step as this will reduce the volume of the extract as well as purify the desired product.

Ion exchange. Ion exchange chromatography is probably the most generally used step in large scale purification. When volumes are very large batch adsorption and also elution can often prove successful. For example, in the purification of L-asparaginase from *Erwinia carotovora* a 6-fold purification and 100-fold volume reduction can be achieved by batch adsorption and elution from CM-cellulose[2]. Ion exchange chromatography, using gradient elution has been used in the purification of a thermostable glycerokinase from 20 kg *Bacillus stearothermophilus*, where a 40 litre column (80 x 25 cm) of DEAE-Sepharose was eluted with a 200 litre linear gradient of increasing phosphate concentration[29]. A larger scale demonstration of the power of this technique is afforded by the simultaneous purification of many enzymes from *B. stearothermophilus* on an 86 litre column (80 x 37 cm) of DEAE-Sephadex eluted with a 400 litre gradient of increasing phosphate concentration[30].

Gel filtration. Gel filtration chromatography is rarely used in the early stages of large-scale enzyme fractionation. Its major early stage application is in group separation, where the species being separated differ considerably in molecular size, such as proteins and salt, ethanol or carbohydrates. This type of

Table 2

Chromatographic methods for the large scale purification of enzymes

Molecular property exploited	Chromatography type	Characteristics	Application
Size	Gel filtration	Resolution moderate for fractionation. Good for buffer exchange. Capacity limited by volume of sample. Speed is slow for fractionation. Fast for buffer exchange.	Fractionation is best left to later stages of a purification. Buffer exchange can be used at any time, although sample volume may be a limitation.
Charge	Ion exchange	Resolution can be high. Capacity is high and not limited by sample volume. Speed can be very high depending on matrix.	Is most effective at early stages in fractionation when large volumes have to be handled. Can also be used in a batch mode.

Table 2 (Cont'd)

Molecular property exploited	Chromatography type	Characteristics	Application
Polarity	Hydrophobic interaction	Resolution is good. Capacity is very high and not limited by sample volume. Speed is high.	Can be applied at any stage, but is most usefully applied when the ionic strength is high, after ion exchange or salt precipitation.
Biological Affinity	Affinity	Resolution can be very high. Capacity can be high, but may be low, depending on ligand. Is not limited by sample volume. Speed is high.	Can be used at any stage, but not normally recommended at an early stage. Batch adsorption is possible.

separation employs highly cross-linked gels which are not readily compressible and which allow high flow rates. Examples include the separation of ethanol from albumin, the separation of insulin from proteolytic enzymes and pro-insulin[31] and the separation of milk-whey proteins from lactose and salt[32]. However, large-scale gel filtration chromatography of enzymes has been applied to the purification of restriction endonucleases using a 40 litre column (80 x 25 cm) of Sephacryl S-200 operated at 2 litres/hr[33] and to the purification of alkaline phosphatase from human plasma on a 20 litre column of Ultrogel AcA 34[34].

Hydrophobic interaction chromatography. Immobilised hydrocarbon chains, most commonly octyl- or phenyl-agarose, can bind many enzymes, particularly under conditions of high ionic strength. For this reason hydrophobic chromatography is best applied when the ionic strength of the extract is high, such as following salt precipitation or ion exchange chromatography. Protein adsorbed onto a hydrophobic matrix can be eluted by a change in pH, reduction in ionic strength or increase in the concentration of chaotropic ions such as thiocyanate, or ethylene glycol[35]. The technique has been successfully applied to the purification of gluco cerebroside-β-glucosidase from human placenta, which was adsorbed on either decyl or octyl agarose columns and eluted with gradients of ethylene glycol[36]. Chromatography on columns of phenyl agarose has been successfully employed in the purification of aryl acylamidase from *Pseudomonas fluorescens*. The enzyme derived from 2 kg bacteria was eluted from an ion exchange column in 0.3 M phosphate buffer pH 7.6 and applied directly to a 500 ml column of phenyl-Sepharose in the same buffer. Elution was affected by a decreasing gradient from 0.1 to 0.01 M Tris HCl pH 7.6[22].

Affinity chromatography. Perhaps the greatest advance in enzyme purification in the past decade has been the introduction of affinity chromatography. On the laboratory scale, this can provide a uniquely powerful method for the rapid isolation of an enzyme; however its use in large scale work has been limited.

There are an enormous variety of ligands and matrices available; for example, nearly a dozen different ATP and NAD agaroses are commercially available. The coupling of labile compounds to a matrix is difficult and time consuming and in many cases the resulting ligand is unstable and subject to biological degradation. However, there are a wide variety of interactions that can be exploited other than those between enzyme and nucleotide cofactor.

Matrices that have been employed include porous glass, polyacrylamide, cross-linked dextran and cellulose, but the most generally useful is agarose. A variety of chemical procedures are available for immobilisation using for example, cyanogen bromide, bis-epoxides, divinyl sulphone, tresyl chloride. All have advantages and disadvantages, but the most frequently employed method is the original procedure utilising cyanogen bromide.

An alternative to natural ligands are the range of reactive dyes produced by ICI and CIBA under the trademarks Procion and Cibacron. These consist of a chromophore, usually an anthraquinone, diazo, or phthalacyanine coupled to a reactive triazinyl chloride group. This group can be used to attach the dyes to the usual range of matrices in a very simple reaction at moderately alkaline pH and ambient temperature. The immobilised dyes are stable and can bind a wide range of proteins and enzymes ranging from albumin to dehydrogenases and kinases[37,38].

For large-scale applications, affinity chromatography is similar to ion exchange in that the matrix can adsorb an enzyme from a large volume either on a column or by batch adsorption. Once bound the desired activity can be eluted by a variety of methods (Table 3). In general terms, biospecific elution can be expected to give the greatest purification but may utilise costly cofactors. Examples of large-scale affinity chromatography are limited and most are in the blood processing field where components of plasma such as thrombin, antithrombin and clotting factors can be isolated on heparin-agarose[39].

Recent summaries of enzymes purified by large-scale affinity chromatography are given by Scawen *et al.*,[40] and Janson and Hedman[31]. It is in this field that the immobilised dyes prove most useful, as they offer superior economics and stability

Table 3

Examples of enzymes purified to homogeneity by large scale dye affinity chromatography

Enzyme	Dye	Eluant	Ref.
Glycerokinase	Procion Blue MX-3G	5mM ATP	29
Glucokinase	Procion Brown H-3R	5mM ATP	this paper
Glycerol dehydrogenase	Procion Red HE-3B	2mM NAD	48
Methionyl tRNA synthetase	Procion Green HE-4BD	Phosphate gradient	49
Tryptophanyl tRNA synthetase	Procion Brown MX-5BR	50mM tryptophan	49
3-hydroxybutyrate dehydrogenase	Procion Red H-3B, Procion Blue MX-4GD	1M KCl 1M KCl + 2mM NADH	41
Malate dehydrogenase	Procion Red H-3B Procion Blue MX-4GD	1M KCl + 2mM NADH 0-700mM KCl gradient	41
Carboxypeptidase G_2	Procion Red H-8BN	Bind in 0.2mM Zn^{2+} pH 7.3 elute with 10mM EDTA pH 5.3 followed with 0.1M Tri HCl pH 7.3	50

compared with their nucleotide cofactor counterparts.

An excellent example of the selectivity of this approach is given by the simultaneous purification of 3-hydroxybutyrate dehydrogenase (HBD) and malate dehydrogenase (MD) from Rhodopseudomonas sphaeroides on two different dye-agarose columns[41]. The crude extract from 1 kg cell paste was loaded directly onto a 1.8 litre column of Procion Red H-3B-Sepharose. The column was washed with buffer to remove unbound protein. HBD was eluted with 1 M KCl and MD eluted with 2 mM NADH in 1 M KCl. These two pools were desalted and loaded separately onto columns of Procion Blue MX-4GD-Sepharose. After washing the columns HBD could be eluted with 1 M KCl containing 2 mM NADH and MD eluted with a gradient of 0 - 700 mM KCl. Both enzymes were obtained in homogeneous form and with a recovery greater than 60%. The efficiency of this purification should be compared with the classical approach which required eight steps and gave a recovery of 9%.

A different immobilised dye has been used to purify glycerokinase from Bacillus stearothermophilus[29]. In this case, the crude extract from 20 kg cells was first chromatographed on a 40 litre column of DEAE Sephadex before the glycerokinase was purified on a column of Procion Blue MX-3G-Sepharose. A 3.5 litre column bound 1 x 10^6 units of enzyme (8 gram) which could be bio-specifically eluted with 5 mM ATP in homogeneous form. A recent development of this process has enabled the simultaneous isolation of a glucokinase from the same extract. After ion exchange chromatography, which only partially separates the two enzymes, the eluate is loaded onto a column of Procion Brown H-3R-Sepharose. The effluent from this column, free of glucokinase but still containing glycerokinase, is applied to the Procion Blue MX-3G Sepharose column as before. The glucokinase can be eluted with 5 mM ATP. Table 3 lists examples of enzymes that have been purified on a large-scale by dye affinity chromatography.

SCALE-UP CONSIDERATIONS

The most important scale-up consideration is the following: If large-scale purification is intended the restrictions imposed by large-scale working should be taken into account at the earliest stages of designing the purification protocol, including the lab. scale development work.

Choice of matrix. The correct choice of matrix is critical if large-scale chromatography is to be successfully applied. A chromatography matrix should consist of rigid,uniform, macro-porous hydrophilic spherical particles. The particles should be chemically inert, insoluble and easily derivatised. No single matrix available conforms to all of these properties, but a wide range of matrices based on both natural and synthetic polymers are now available. The traditional matrices, based on dextran, agarose and cellulose suffer from insufficient rigidity at porosities suitable for protein chromatography. The newer, macroporous matrices offer a much more rigid structure, and are either derivatives of natural products like the cross-linked agaroses (Sepharose-CL) or entirely synthetic like Trisacryl or Spheron. These newer matrices are suitable for all types of chromatography, but their improved hydrodynamic properties make them ideally suited for large-scale applications when high flow rates are desirable. These properties lead to other advantages since such matrices do not show appreciable variations in volume with changes in ionic strength or pH and can be regenerated using NaOH solutions without being removed from the column.

Choice of column. The choice of column is largely restricted to increasing the cross-sectional area in proportion to the increase in sample volume. The bed height and linear flow rate should be kept constant, although it may be possible to increase the flow rate in order to increase sample through-put. The design of the column should conform to the standards of laboratory scale columns in terms of dead volumes and should be compatible with the process solvents and samples. For ion exchange and affinity chromatography short, broad columns are desirable. For fractionation by gel filtration longer columns are necessary; the 'Stack' column system offers some advantages when soft gels are used, as the short sections reduce the compression forces acting on the gel. Detailed consideration of the desirable features of large-scale chromatography columns is given by

Janson and Hedman[31].

Ultrafiltration. Ultrafiltration has become a standard laboratory technique for the concentration of protein solutions, for example, following gel filtration chromatography. It can also be used as an alternative to dialysis or gel filtration for desalting or buffer exchange. By using affinity precipitants to increase the molecular weight of the desired protein it can also be used as a method of purification. Hollow fibre ultrafiltration units for pilot scale use are available with up to 6.4 m^2 of membrane area, giving ultrafiltration flow rates up to 200 litres per hour, depending on protein concentration. Larger units are available with ultrafiltration flow rates of several hundred litres per hour, making this method applicable on almost any scale.

Aqueous two-phase separation. Aqueous two-phase systems can be created by mixing solutions of polyethylene glycol and dextran or polyethylene glycol and salts, such as ammonium sulphate or potassium phosphate. Proteins and cell debris partition between the two phases. The exact location of a particular protein depends on such parameters as its molecular weight, the concentrations and molecular weights of the polymers, the temperature, pH and ionic strength of the mixture and the presence of polyvalent salts[42]. The optimum conditions for favourable partition have to be found empirically for each protein. Two-phase systems can be effectively applied to remove debris from cell homogenates and at the same time achieving a degree of purification.

The two phases can, in some cases, be separated in a settling tank, but a more rapid and efficient separation can be achieved by using centrifugal separation. This technique can be used to great advantage in large-scale enzyme preparations[5]. However, despite these apparent advantages aqueous phase partition is only rarely used, perhaps because of the non-recoverable cost of the dextran and polyethylene glycol employed, although a recent report suggests that it is possible to use crude dextran with a considerable saving in cost[43].

Two phase separation has been used for the separation and large-scale purification of pullulanase and 1,4-α-glucan phosphorylase from 5 kg Klebsiella pneumoniae cell paste[44], and RNA polymerase and glutamine synthetase from E. coli[45]. By coupling a ligand to polyethylene glycol it is possible to increase the partition coefficient of a protein in favour of the polyethylene glycol phase. Thus, by using the dye Cibacron blue it was possible to achieve a 58-fold purification of yeast phosphofructokinase in two steps[46].

CONCLUSION AND THE FUTURE

In this paper we have discussed the major techniques currently available for the large-scale isolation of enzymes. It is important to remember that in many cases it is unnecessary to purify an enzyme to homogeneity; it is sufficient for many diagnostic reagents only to remove interfering activities. However, therapeutic proteins must be homogeneous and free of non-protein contaminant such as pyrogens.

In future, it is likely that other techniques for protein purification will become available. One obvious example is HPLC, already a widely established analytical method, which is increasing in its scale of application. Perhaps the most promising use of HPLC is in high performance liquid affinity chromatography, which combine the benefits of HPLC with the biological specificity of affinity ligands. However, there are still problems over ligand immobilisation and matrices to be overcome before the method can become widely applicable. Also available are techniques like affinity precipitation which, when linked with ultrafiltration, may become a rapid and elegant method of enzyme purification.

The formation of dense cytoplasmic inclusion granules within E. coli provides a means of rapidly purifying foreign proteins produced by rDNA technology. Such granules have been isolated by a single step centrifugation for both prochymosin and bovine growth hormone produced in E. coli. Specific washing of the granules to release contaminants frequently yields insoluble material with a purity of over 50% and sometimes near homogeneity.

Solubilisation and renaturation of such granules requires conditions specific to the stability of the particular protein following which such proteins can be purified by conventional means.

Recombinant DNA technology per se has also been used to assist in the purification of proteins. C-terminal genetic fusion of a polyarginine tail to urogastrone has illustrated a novel approach to purification in which, due to the unusual basicity of the fused protein, it is strongly bound by a cationic ion exchange matrix[47]. Since the latter group of matrices only bind 10% of cellular proteins under most conditions a large purification is obtained. The polyarginine tail can be removed by immobilised carboxypeptidase B and rechromatography of the liberated recombinant protein on the same matrix results in a major change in its elution profile, but not in that of the contaminants. An alternative approach is that of fusion of the protein A gene to the gene for the desired recombinant product. This provides a mechanism by which the latter may be purified by virtue of the immunoglobulin binding ability of protein A (Uhlen et al., 1983, 1984). This latter technique however has two disadvantages, firstly the extreme conditions of pH or chaotropic agent required to elute protein A from IgG-matrices, which may denature the required product, and secondly the difficulty of removing the protein A moiety without disruption of the structure of the required recombinant product.

It is likely that in the future, as now, protein purification will be based on a set of generally applicable techniques, augmented by a few specialised techniques which are only applicable to specific proteins, but which can result in spectacular purifications.

REFERENCES

[1]K. Anstrup, In "Applied Biochemistry and Bioengineering", (Eds. L.B. Wingard, E. Katchalski-Kalizin and L. Goldstein), Academic Press, London and New York, 1979, Vol. 2, p. 27-56.

[2]H.E. Wade, British Patent Specification B1258068, 1968.

[3]P.J. Halling, J.A. Asenjo and P. Dunnill, Biotech. Bioeng., 1979, 21, 2359-2363.

[4]B.C. Buckland, W. Richmond, P. Dunnill, P. and M.D. Lilly, In "Industrial Aspects of Biochemistry", (Ed. B. Spencer), North Holland, Amsterdam, 1974, 65-79.

[5]K.H. Kroner, H. Hustedt and M.R. Kula, Biotech. Bioeng., 1982, 24, 1015-1045.

[6]H.C. Neu and L.A. Heppel, J. Biol. Chem., 1965, 240, 3685-3692.

[7]S.E. Charm and C.C. Matteo, Methods in Enzymology, 1971, 22, 476-556.

[8]M.I. Haas and J.E. Dowding, Methods in Enzymology, 1975, 43, 611-628.

[9]Gunsalas, A. Miguel, E.A. Meighen, M.Z. Nicoli and K.H. Nealson, J. Biol. Chem., 1972, 247, 398-404.

[10]L. Edebo, In "Fermentation Advances", (Ed. D. Perlman), Academic Press, London and New York, 1969, 249-271.

[11]K.E. Magnusson and L. Edebo, Biotech. Bioeng., 1976, 18, 975-986.

[12]J. Melling, C.G.T. Evans, R. Harris-Smith and J.E.D. Stratton, J. Gen. Microbiol., 1973, 77, XVIII.

[13]F. Marfy and M.R. Kula, Biotech. Bioeng., 1974, 16, 623-634.

[14]J. Rehacek and J. Schaeffer, Biotech. Bioeng., 1977, 19, 1523-1534.

[15]J. Woodrow and A.V. Quirk, Enzyme and Microbial Technology, 1982, 4, 385-389.

[16]J. Darbyshire, In "Topics in Enzyme and Fermentation Biotechnology", (Ed. A. Wiseman), Ellis Horwood, Chichester, UK, 1981, 5, 145-186.

[17]M. Follows, P.J. Hetherington, P. Durrill and M. Lilly, Biotech. Bioeng., 1971, 13, 549-560.

[18]A. Atkinson, Process Biochemistry, 1973, 8, 9-13.

[19]J. Melling and G.K. Scott, Biochem. J., 1972, 130, 55-66.

[20]J. Melling and A. Atkinson, J. Appl. Chem. Biotechnol., 1972, 22, 739-744.

[21]A.V. Quirk, and J. Woodrow, Biotech. Letts., 1983, 5, 277-282.

[22]P.M. Hammond, C.P. Price and M.D. Scawen, Eur. J. Biochem., 1983, 132, 651-655.

[23]J.K. Baird, R.F. Sherwood, R.J.G. Carr and A. Atkinson, FEBS Letts., 1976, 70, 61-66.

[24]M.S. Le, P.S. Ward and T. Atkinson, Biotech., 1984, 84, II A97-A116.

[25]M. Dixon and E.C. Webb, "Enzymes", Longmans, London, 1979.

[26]P. Foster and J.G. Watt, In "Methods of Plasma Fractionation", (Ed. J. Curling), Academic Press, London and New York, 1980, 17-31.

[27]M.R. Kula, W. Honig and H. Foellmer, In "Proceedings of International Workshop on Technology for Protein Separation and Improvement of Blood Plasma Fractionation", (Ed. H.E. Sandberg), NIH Pub. No. 78-1422, Washington DC, USA, 1977, 361-371.

[28]Y.L. Hao, K.C. Ingham and M. Wickerhauser, In "Methods of Plasma Protein Fractionation", (Ed. J. Curling),Academic Press, London and New York, 1980, 57-74.

[29]M.D. Scawen, P.M. Hammond, M.J. Comer and T. Atkinson, Anal. Biochem., 1983, 132, 413-417.

[30]T. Atkinson, G.T. Banks, C.J. Bruton, M.J. Comer, R. Jakes, T. Kamalagharan, A.R. Whitaker and G.P. Winter, J. Appl. Biochem., 1, 247-258.

[31]J.C. Janson and G. Hedman, Adv. Biochem. Eng., 1982, 25, 43-99.

[32]R.H. Delaney, In Applied Protein Chemistry, (Ed. R.A. Grant), Applied Science Publishers, Barking, Essex, UK, 1980, 233-280.

[33]A.H.A. Bingham and T. Atkinson, Biochem. Soc. Trans., 1978, 6, 315-324.

[34]R. Hanford, W. d'A. Maycock and L. Vallet, In "Chromatography of Synthetic and Biological Polymers", (Ed. R. Epton), Ellis Horwood, Chichester, U.K., 1978, Vol. 2, 111-119.

[35]S. Hjerten, J. Rosengren and S. Pahlman, J. Chromatog., 1974, 101, 281-288.

[36]F.S. Furbish, H.E. Blair, J. Shiloach, P.G. Pentchev and R.O. Brady, J. Biol. Chem., 1977, 74, 3560-3563.

[37]C.R. Lowe, D.A.P. Small and T. Atkinson, Int. J. Biochem., 1981, 13, 33-40.

[38]G. Kopperschlager, H.-J. Bohme and E. Hofman, Adv. Biochem. Eng., 1982, 25, 101-138.

[39]E.A. Hill and M.D. Hirtenstein, In "Advances in Biotechnological Processes", (Eds. A. Mizrahi and A. van Wezel), Alan R. Liss, Inc., New York, 1983, 31-66.

[40]M.D. Scawen, A. Atkinson and J. Darbyshire, In "Applied Protein Chemistry", (Ed. R.A. Grant), Applied Science Publishers, Barking, Essex, UK, 1980, 281-324.

[41]M.D. Scawen, J. Darbyshire, M.J. Harvey and T. Atkinson, Biochem. J., 1982, 203, 699-705.

[42]M.R. Kula, In "Applied Biochemistry and Bioengineering", (Eds. L.B. Wingard, E. Katchalski-Katzir and L. Goldstein), Academic Press, London and New York, 1979, Vol. 2.

[43]K.H. Kroner, H. Hustedt and M.R. Kula, Biotech. Bioeng., 1982, 24, 1015-1045.

[44]H. Hustedt, K.H. Kroner, W. Stach and M.R. Kula, Biotech. Bioeng., 1978, 20, 1989-2005.

[45]T. Takahashi and Y. Adachi, J. Biochem. (Tokyo), 1982, 91, 1719-1724.

[46]G. Kopperschlager and G. Johansson, Anal. Biochem., 1982, 124, 117-124.

[47]H.M. Sassenfeld and S.J. Brewer, Biotechnology, 1984, 2, 76-81.

[48]M.D. Scawen, P.M. Hammond and K.J. Bown, unpublished data, 1984.

[49]C.J. Bruton and T. Atkinson, Nucleic Acids Res., 1979, 7,

[50]R.F. Sherwood, R. Melton, S. Alwan and P. Hughes, Eur. J. Biochem., in press, 1985.

13
Biosensors

By M. Gronow, C. F. M. Kingdon, and D. J. Anderton

CAMBRIDGE LIFE SCIENCES P.L.C., CAMBRIDGE SCIENCE PARK, MILTON ROAD, CAMBRIDGE CB4 4BH, U.K.

1. Introduction

The early 1980's have seen rapid advances in the application of molecular biology and other techniques of the life sciences to create new commercial products and processes which we now call biotechnology. While genetic engineering and monoclonal antibody technology are prominent leaders in biotechnology at present, the coupling of immobilized enzymes and other biological materials to micro-electronics (a technology itself evolving at an incredible rate) promises to be another fertile area of the biotechnology revolution. By these combinations of biochemistry and microelectronics BIOSENSORS are being produced. When fully developed they will form the basis of many analytical systems with such varied applications as health care and industrial monitoring. In the future it is hoped that the information gained from the construction of miniaturized biosensors will lead to the discovery of workable molecular switching devices - true biochips, which can be used for computer construction.

The purpose of this chapter is to present an overview of the field of biosensors. It is not intended to consider each type of device in detail but to give the reader sufficient appetite and interest to seek further information.

What are biosensors? Older definitions included any probe which placed in a biological media gave a quantifiable signal. This of course included pH and ion selective electrodes but these could probably now be described merely as biological probes (bioprobes). The debate may be continued depending on whether you regard a biosensor as a sensor which detects biological materials or a probe containing a specific biological material which can be used to detect biochemicals or chemicals in general. In an attempt to simplify the situation we chose the latter description which fits more closely with biotechnological concepts. Therefore, we came to the following:-

"A biosensor is an analytical tool or system consisting of an immobilised biological material (such as an enzyme, antibody, whole cell, organelle or combinations thereof) in intimate contact with a suitable transducer device which will convert the biochemical signal into a quantifiable electrical signal."[1,2]

Sticking broadly to this definition we can evolve the schematic shown in Figure 1.

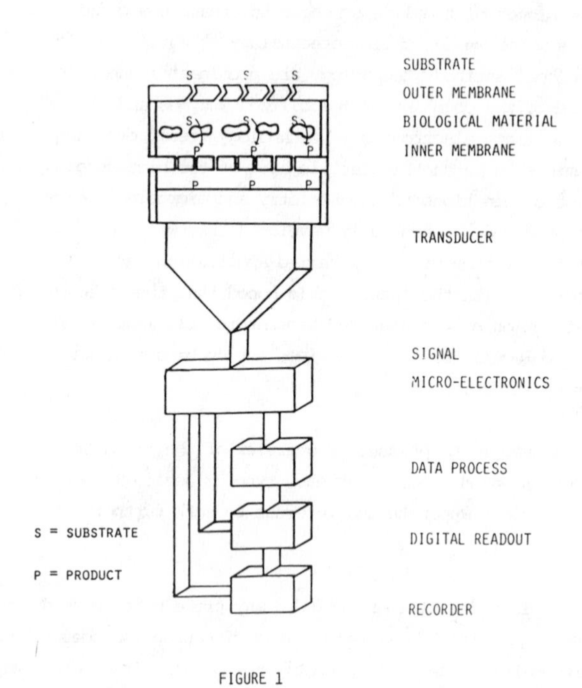

FIGURE 1

In some ways one might say that any of the increasingly sophisticated analytical equipment used for bioanalysis that has evolved over the last 20 to 30 years could form part of a biosensor depending on your definition of the transducer, for example UV and IR spectrophotometers, fluorimeters, NMR equipment etc. However, we feel that biosensors essentially differ from existing techniques in perhaps three useful and fundamental ways, i.e.

i. In the intimate contact of the biological material, (whether it be whole cells, organelles, antibodies or enzymes) with a transducer which converts the biological signal into a quantifiable electrical signal.

ii. In its functional size. The 'sensing' portion of the biosensor is generally small and this allows small test sample size, minimum interference with existing processes following implantation, and access to difficult and possibly dangerous environments, without interrupting process flow.

iii. The biological material can be selected to suit the analytical needs and operate at various levels of specificity. It may be highly specific, specific to a narrow band of compounds or exhibit broad spectrum specificity. A theoretical illustration of this graduation in specificity would be a sensor sensitive to a single antibiotic, for example gentamycin, or to all aminoglycosides or all antibiotics. This flexibility in choice of biological material allows the user to tailor the biosensor to the required need.

2. Immobilization - the Key to Biosensor Construction

The last 15 years have seen great advances in the technology of immobilizing biological materials to various natural and manmade materials, such as membranes and gel matrices. The immobilization techniques available can be summarized as follows:

i. ADSORPTION - the oldest technique. The biological material is simply contacted with a surface active material such as charcoal, alumina, clay, glass or ion-exchange resins. A major disadvantage is that desorption can occur simply by variations in pH, ionic strength, temperature, solvent, etc. This technique is often only suitable for a one-off measurement.

ii. COVALENT ATTACHMENT - commonly used. Chemical groupings on the biological material not essential for activity (e.g. non-essential amino acid residues of enzymes) are attached to chemically activated supports such as synthetic polymers, cellulose, glass, etc. The covalent bonds formed provide the stable linkage.

iii. CROSS-LINKING. Intermolecular covalent linkages can be formed in macromolecules with the aid of bifunctional cross-linking agents (e.g. diisothiocyanates, alkylating agents, dialdehydes). A major disadvantage is that some systems lose biological activity when their three-dimensional structure is altered in this way.

iv. ENTRAPMENT - one of the most popular techniques. Gels and polymers can be used. Materials such as polyacrylamides, silica gel and starch are cross-linked in the presence of the enzyme or other biological material which is thereby trapped or captured. The pores of the matrix are large enough to allow substrate and product to diffuse through but small enough to prevent loss of the active high molecular weight biological material.

v. MICROENCAPSULATION. Materials are trapped by the membranes of various polymers, usually in capsules with mean diameters ranging from 5-300μm. The entrapped biological material is too large to move though the pores of the semi-permeable synthetic membrane, but smaller substrate and product molecules can move back and forth through the pores of the membrane readily.

The immobilization techniques for enzymes have been reviewed by Zaborsky[3] and Chibata[4]; whole cells[5] and organelles[6] have also received considerable attention. It should be noted that the preparation of highly active and stable materials for use in large-scale biotechnology has often established the groundwork for the successful use of immobilized biologicals in analytical devices.

3. Construction of Biosensor Devices

Many forms of biosensor devices have been described and others are clearly being created at an ever increasing rate. Table 1 illustrates the variety of combinations now available.

TABLE 1

Tranducer	Biological Material Intimately Associated with the Tranducer
Conductimetric	Enzymes, antibodies
Fibre Optic Devices	Enzymes, antibodies
FET's	Enzymes and/or antibodies
Gas Sensing Electrodes	Enzymes, whole cells, organelles
Ion Selective Electrodes	Enzymes
Optical Devices, eg Optoelectronic	Enzymes, antibodies
Oxygen Electrode	Enzymes and/or antibodies, whole cells, organelles
pH Electrode	Enzymes, whole cells, organelles
Photodiodes	Enzymes and/or antibodies
Piezoelectric Crystals	Whole cells
Semiconductors/Conductors	Enzymes, whole cells
Thermistor Devices	Enzymes, antibodies, whole cells

BIOSENSOR COMBINATIONS

4. Biosensor Development

The earliest biosensors consisted of immobilized enzymes in conjunction with pH or oxygen electrodes as constructed by Clark and Lyons[7] and Updike and Hicks[8] in the 1960's. A generalized form of this type of biosensor is illustrated in Figure 2.

The evolution of these electrochemical type of biosensors has been extensively reviewed by Carr and Bowers[9] and many others (e.g. 10, 11, 12, 13, 14). Figure 3 illustrates the variety of biosensor combinations which arose from electrochemical concepts in the 1970's.

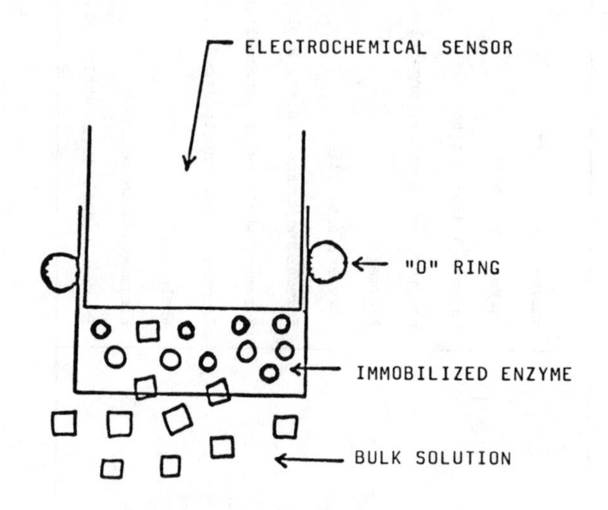

□ SUBSTRATE TO BE MEASURED

○ INERT PRODUCT

○ ELECTRODE MEASURABLE PRODUCT

FIGURE 2.

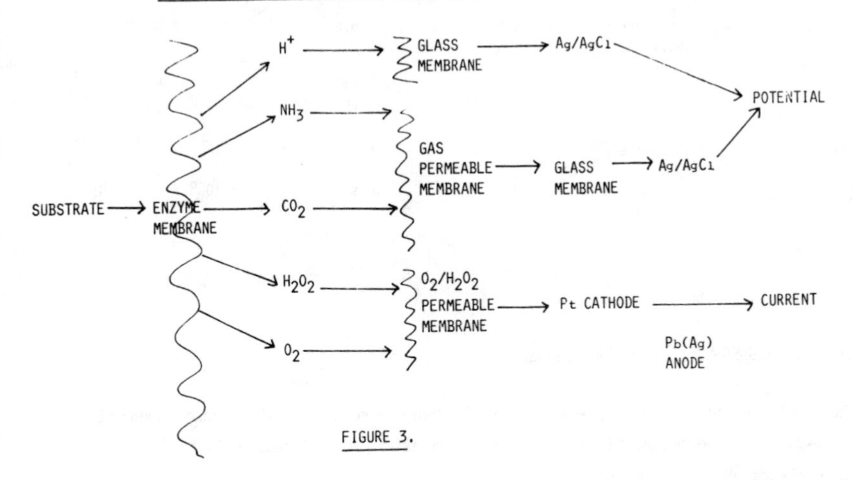

FIGURE 3.

Generally devices consisted of either the potentiometric type, in which a potential was measured (a logarithmic response according to the classic Nernst equation), or the amperometric type in which current changes were monitored. Over 500 papers have been published on these electrochemical types of biosensor but about 45% of these have been concerned with glucose or urea measurement (using the cheap and readily available glucose oxidase and urease enzymes). In general, most simple enzyme electrode biosensors exhibit a linear response in the range of 10^{-2} to 10^{-4} M of substrate while some electrodes respond to concentrations as low as 5×10^{-7} M and/or as high as 10^{-1} M. This work has been extensively tabulated (e.g. see 7) but in Table 2 we give some selective examples of this type of biosensor. Generally the life-time of such biosensors is long, at least 3 - 4 weeks (sometimes several months) and the reponse is rapid (less than 60s).

TYPICAL ENZYME BIOSENSORS (SCOPE)

SUBSTRATE	ENZYME	STABILITY	RESPONSE TIME	RANGE
ALCOHOL	ALCOHOL OXIDASE	120 DAYS	30 SEC	$5\text{-}10^3$ MG ML^{-1}
CHOLESTEROL	CHOLESTEROL OXIDASE/ CH. ESTERASE	30 DAYS	2 MIN	$10^{-2} - 3 \times 10^{-5}M$
URIC ACID	URICASE	120 DAYS	30 SEC	$5 \times 10^{-3} - 5 \times 10^{-5}M$
SUCROSE	INVERTASE	14 DAYS	6 MIN	$2 \times 10^{-3} - 10^{-2}M$
GLUCOSE	GLUCOSE OXIDASE	50 - 100 DAYS	10 SEC	$3 \times 10^{-6} - 2 \times 10^{-3}M$

TABLE 2

5. Biosensor Diversification

Recently we have seen the evolution of three more types of electrochemical biosensor; the conductimetric, redox mediated[15] and field effect transistor FET[16] devices.

i. CONDUCTIMETRIC BIOSENSORS

The conductimetric system employs two pairs of identical small conductivity electrodes in a flat configuration. Across one pair of electrodes is positioned a membrane to which enzyme has been immobilised, whilst over the second pair is a 'blank' membrane with no enzyme present. The instrumentation measures conductivity across each pair of electrodes in turn at a set frequency. In the presence of the enzymes' substrate, localised conductivity changes can be monitored in the proximity of the enzyme-linked membrane which are dependent upon substrate concentration. By monitoring the difference in response observed at the two pairs of electrodes, the 'blank' pair can be used as a reference to allow back-off of the inherent background conductivities of biological sample. Figure 4 illustrates the range of analytes successfully measured by this simple biosenor.

ENZYME CONDUCTIMETER

UREA	**urease**
PENICILLIN G	**penicillinase**
AMPICILLIN	**penicillinase**
CREATINE	**creatinase**
ASPARAGINE	**asparaginase**
CREATININE	**creatinine amidohydrolase**
PROTEIN	**pronase**
GLUCOSE	**glucose oxidase**
PARACETAMOL	**arylacylamide amidohydrolase**

A selection of immobilized enzymes used in the construction of working conductimetric electrodes

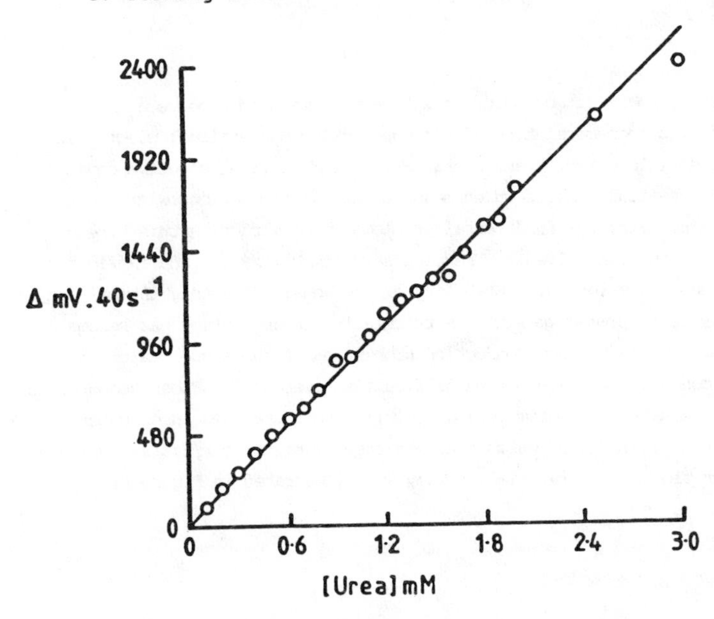

The enzyme conductimeter: a typical calibration curve for urea determined with a urease conductimeter

FIGURE 4.

ii. REDOX-MEDIATED BIOSENSORS

The redox-mediated concept biosensor arose from fundamental work being carried out on bio-fuel cells. The aim in constructing this type of biosensor is to facilitate the transfer of electrons generated by an oxido-reductase enzyme (or enzyme system) to the electrode surface. Natural mediators such as the cytochromes have been shown to promote passage of electrons but one of the most recent electron mediators showing great promise is ferrocene and its derivatives[15]. The principle of this type of biosensor using other redox mediators and some data showing the response that can be obtained in our laboratories with glucose is shown in Figure 5. The response time of this type of biosensor can be extremely rapid - in the order of seconds. With the development of more efficient organic semiconductors (usually by "doping" methods) one can expect to see an even more intimate association between the enzyme employed and the electrode surface leading to extensive miniaturization possibilites.

iii. FETs

Over the last 10 years strenuous efforts have been made to produce a miniaturized electrochemical type of biosensor using conventional electronic devices called field effect transistors (FETs), ISFETs (ion selective devices) or CHEMFET's (chemical sensors which measure energy of reactions with whole molecules). However, the fundamental problems in constructing this type of biosensor have not been solved. The necessary immobilization and fabrication technologies need developing. Stability of the sensing element, thermal and chemical, has to be investigated. Particularly, encapsulation has become a major problem and the surface conduction properties of the sensor material, such as silicon nitride, have proved difficult to overcome. These sensor-like "chips" (approximately 30μ diameter) are similar to those used in computers except that the metal "gate" which controls the transistor current is replaced by organic or biochemical material. They are illustrated in Figure 6.

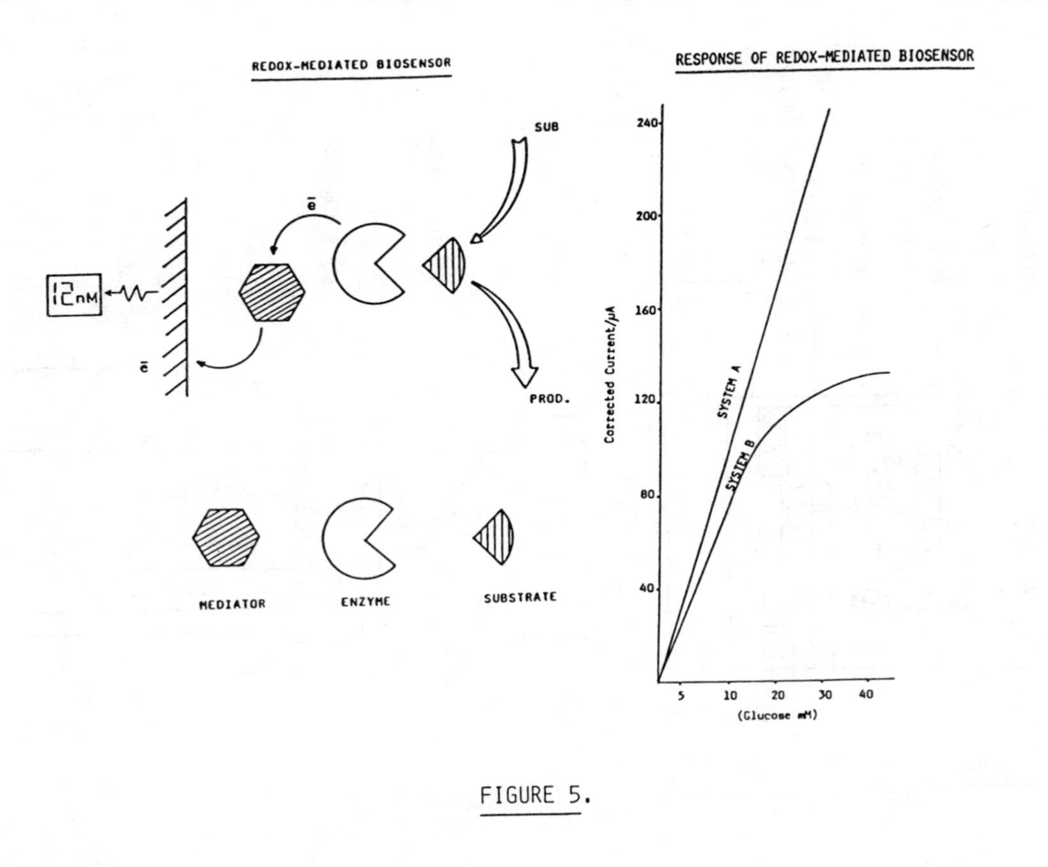
REDOX-MEDIATED BIOSENSOR
12nM
SUB
PROD.
MEDIATOR
ENZYME
SUBSTRATE
RESPONSE OF REDOX-MEDIATED BIOSENSOR
SYSTEM A
SYSTEM B
Corrected Current/μA
240
200
160
120
80
40
5
10
20
30
40
(Glucose mM)

FIGURE 5.

SCHEMATIC OF A FET BIOSENSOR

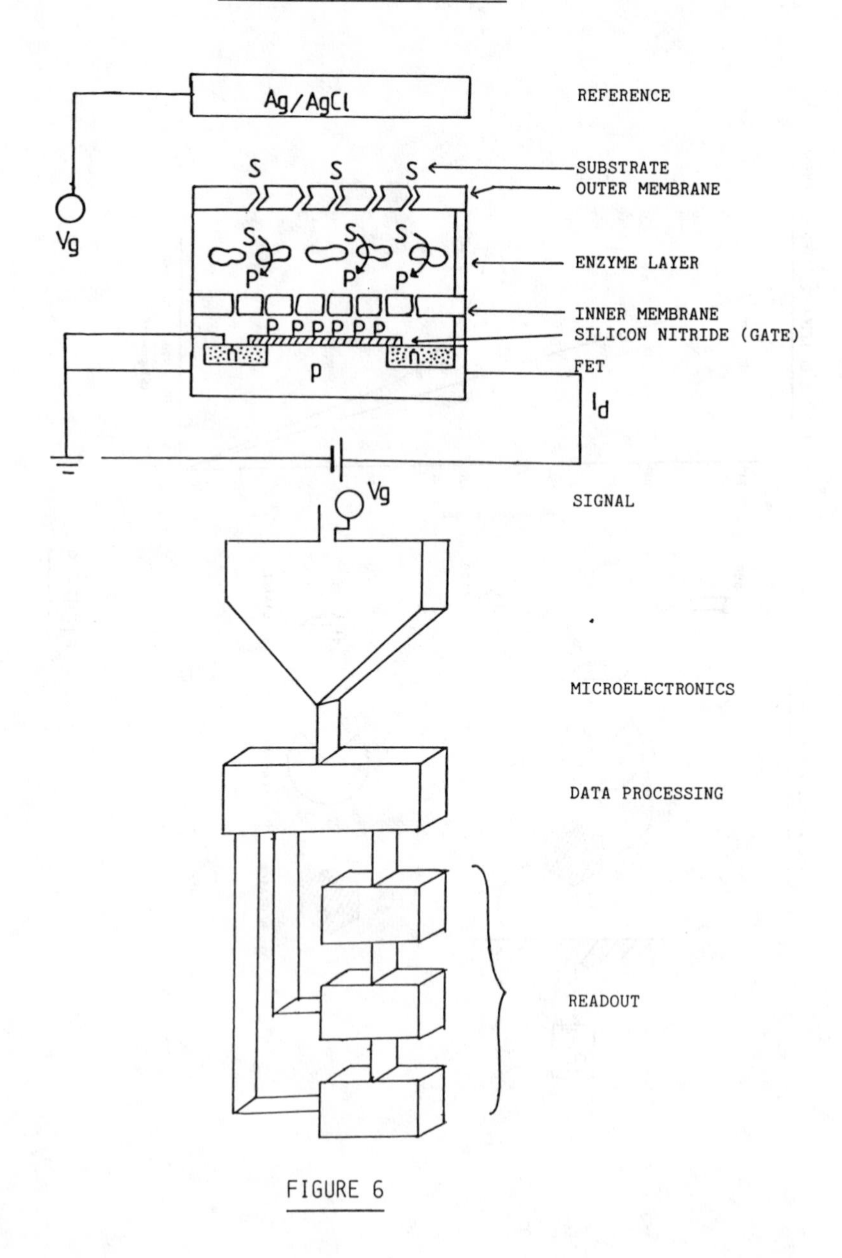

FIGURE 6

The sensing material responds to the change in environment, either gas or liquid. The response has a field effect on the source to drain current off the field effect transistor. Usually this current is held at a fixed valve whilst the voltage necessary from the gate to maintain it is monitored.

iv. THERMISTOR TYPE

An interesting form of biosensor evolved in the 1970's. It utilized a thermistor device which was able to monitor the small temperature differences that were generated in biochemical reactions. A linear response in temperature in the range 0.01 to 0.001 ^{o}C is often obtained. Thermal enzyme analysis was pioneered by American and Swedish groups in the form of probe and flow-through devices, but the miniaturization of the device is still essential to ensure an acceptable biosensor format. Some examples of assays performed with the enzyme thermistor by Swedish workers[17] are shown in Table 3 showing that in general this sensor is more sensitive than the electrochemical variety.

TABLE 3

EXAMPLES OF ASSAYS PERFORMED WITH THE ENZYME THERMISTOR

SUBSTANCE	ENZYME(S)	CONCENTRATION RANGE x 10^{-6}M
ASCORBIC ACID	ASCORBIC ACID OXIDASE	50 - 100
CHOLESTEROL	CHOLESTEROL OXIDASE	30 - 150
GLUCOSE	GLUCOSE OXIDASE + CATELASE	0.2 - 700
PHENOL	TYROSINASE	100 - 1000
SUCROSE	INVERTASE	50 - 100,000

v. OPTOELECTRONIC

Another novel form of biosensor using optical principles was developed around 1980 by Lowe and his collaborators[18] and was called the opto-electronic sensor. This is represented diagramatically as a flow-through device in Figure 7.

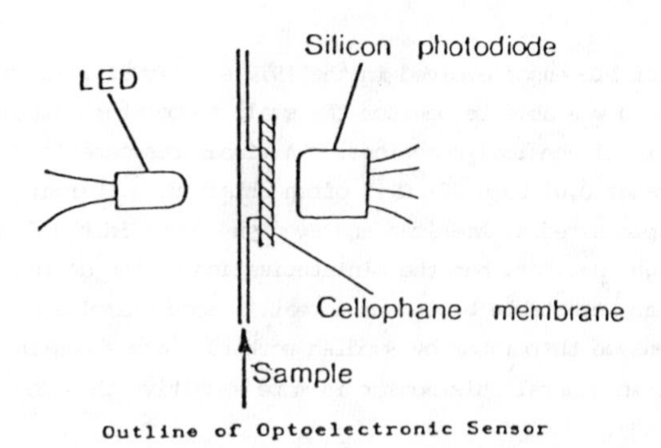

Outline of Optoelectronic Sensor

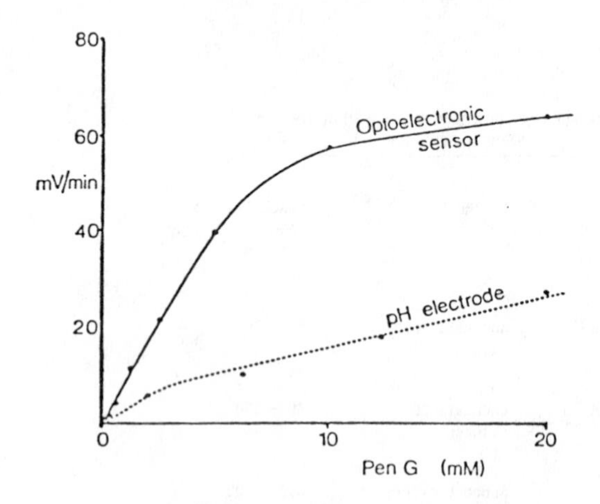

Standard curve of penicillin concentration against photodiode output voltage for the penicillin electrode and the penicillin optoelectronic sensor

FIGURE 7

The immobilized biological element is an enzyme linked to a dye which is in turn linked to a membrane. A pH change generated by the enzyme changes the colour of the dye-membrane complex. The transducer system is a simple, light emitting diode (LED) with a wavelength corresponding to the λ max of the dye and a photodiode. The flow cell illustrated was extremely stable and gave a very acceptable signal. Some characteristics of optoelectronic sensors constructed along these principles are given in Table 4.

TABLE 4

CHARACTERISTICS OF OPTOELECTRONIC SENSORS

SUBSTRATE	RESPONSE RANGE (MM)	VOLTAGE CHANGE Δ MV MIN^{-1}	STABILITY
ALBUMIN	0.07 - 0.5	-	$\gg$ 1 YEAR
UREA	0 - 40	125	$\sim$1 MONTH
D-GLUCOSE	0 - 70	1.5	$\sim$1 WEEK
PENICILLIN-G	0.3 - 5.0	45.5	> 1 YEAR
AMPICILLIN	0 - 10	24.1	

6. Variations on the Biological/Biochemical Components

i. BIOAFFINITY PRINCIPLES

In addition to the classical form of enzyme biosensors described previously a new form of biosensor has arisen which could be called a BIOAFFINITY SENSOR. This is represented diagramatically in Figure 8. An interesting form of this type of biosensor mode was by Aizawa's group in 1983.[19] They used the biotin avidin system in conjunction with an electrode system. Biotin concentrations of 10^{-5} to 10^{-8} g mL^{-1} could be measured in one minute. Other possibilities for biosensor construction using biosubstances capable of molecular recognition are:-

Lectin - for saccharide
Hormone receptors - hormones
Drug receptors - drugs and active drug metabolites
Antibodies - antigens
Nucleic acids (DNA, RNA) - complementary steroids (hybridization)

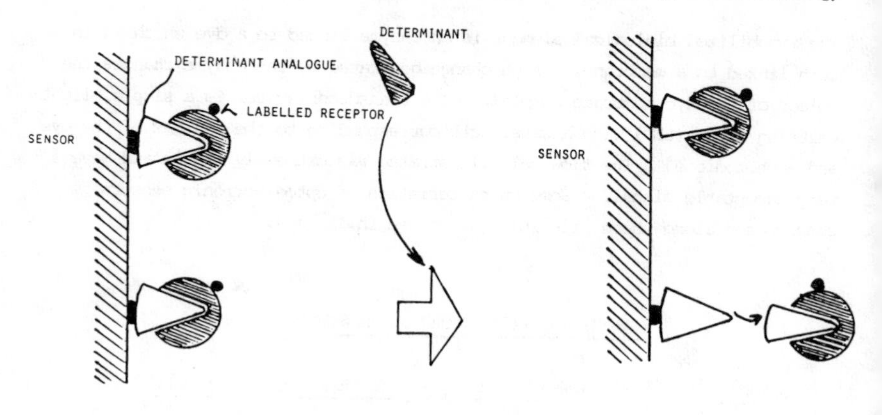

PRINCIPLE OF BIOAFFINITY SENSOR

FIGURE 8

The advantages of bioaffinity sensors can be listed as follows:

a) The concentration measurement is based on equilibrim binding not rate measurement. This makes them less sensitive to local concentrations in micro environment.

b) Sometimes they neither consume the measured compound nor produce a product.

c) Can be miniaturized for implantation.

d) A high degree of selectivity is possible using appropriate receptors.

An example of a very exciting new development by Mansouri & Schultz[20] in this form of biosensor, for glucose measurement, is illustrated in Figure 9.

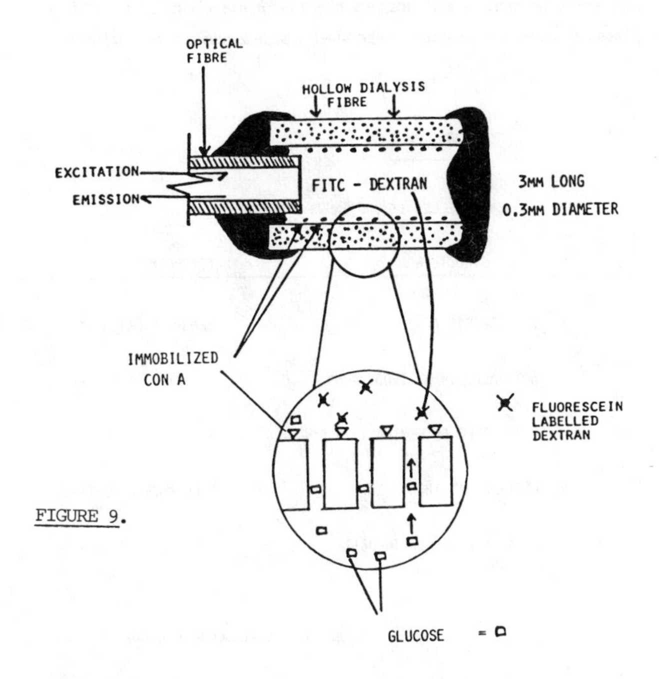

FIGURE 9.

The sensing element consists of a short length of hollow dialysis fibre remotely connected to a fluorimetry instrument via a single optical fibre. This sensing element contains a carbohydrate receptor Concanavalin A, immobilized on its inner surface and a high molecular weight fluorescein labelled dextran as a competing ligand. Glucose in the external medium diffuses through the dialysis fibre into the sensing element and competes with dextran for binding to Con A. At equilibrim, the level of free fluorescein in the hollow fibre lumen is measured via the optical fibre and is correlated to the concentration of glucose. Blood glucose was measured in 5 to 7 minutes and the response of the biosensor was linear from 50 - 400 milligrams percent.

Of course, the potentially most important type of bioaffinity biosensor is that which could be used for immuno-assay. The construction of a successful, simple form of this type of biosensor would open up a vast new range of analytical possibilities. In its simplest form this would involve the antibody, antigen or hapten linked to a pre-conditioned support carrier such as a membrane. A direct change in the transduction principle, e.g. potential or conductivity, can then be monitored across the membrane when interaction occurs. Examples of such simple non-labelled immunosensors are given in Table 5.

TABLE 5.

NON-LABELLED IMMUNOSENSOR

SENSOR	STRUCTURE	MEASUREMENT
SYPHILIS	CARDIORIPIN/ACM*	MEMBRANE POTENTIAL
ALBUMIN	ANTI-ALBUMIN ANTIBODY/ACM*	" "
BLOOD-TYPE	BLOOD-TYPE DETERMINANTS/ACM*	" "
hCG	ANTI-hCG ANTIBODY/TiO_2	ELECTRODE POTENTIAL
ANTIBODY	ANTIGEN-BOUND ERYTHROCITE (LIPOSOMES)	COMPLEMENT BINDING

*ACM: ACETYLCELLULOSE MEMBRANE

The strength of the signal generated will depend on several factors but is dependent on the amount of protein that can be coupled to the surface to give a high density active layer - rather difficult to accomplish in practice.

A more successful form of immunosensor described in the literature is based on enzyme linked immunoassay (EIA or ELISA),. This is represented diagramatically in Figure 10. Table 6 illustrates some examples of EIA published in current journals (e.g. see refs. 21,22,23). Most of the work has been carried out by Japanese workers using electrode systems.

SCHEMATIC OF AN IMMUNOSENSOR - ENZYME LABELLED COMPETITIVE

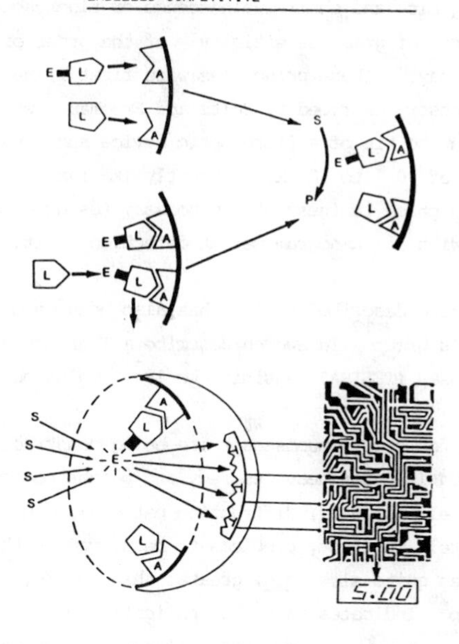

FIGURE 10.

TABLE 6

EXAMPLES OF EIA BASED BIOSENSORS

Sensor	Receptor (label)	Measurement	Range (g/cm^3)
IgG	anti-IgG (catalase)	competition, EIA	10^{-4} - 10^{-3}
	" (")	sandwich, EIA	10^{-6} - 10^{-3}
	" (glucose oxidase)	competition, EIA	10^{-6} - 10^{-3}
IgM	anti-IgM (catalase)	sandwich, EIA	10^{-7} - 10^{-4}
Albumin	anti-Albumin (")	"	10^{-6} - 10^{-3}
hCG	anti-hCG (")	competition, EIA	10^{-2} - 10^{2} IU/cm^3
AFP	anti-AFP (")	"	10^{-11} - 10^{-8}
HBs	anti-HBs (peroxidase)	"	10^{-7} - 10^{-5}

However, as shown in Figure 10 other options exist. These include the use of optical systems employing bio- and chemi-luminescence or fluorescence labels. This approach has the merit of great sensitivity - of the order of that achieved by radioimmunoassay. These principles were first illustrated by a simple bioluminescence sensor described by Seitz and Freeman[24] who immobilized luciferase on the tip of a fibre optic device and managed to monitor ATP in the range of 10^{-6} to 10^{-7} M. Recently Ikariyama et al.[25] described a solid phase luminescent immunoassay (using catalase) for human serum albumin with a 1 nanogram per mL detection limit.

The enzyme thermistor device described earlier has also been used for antibody antigen study. Mosbach's group[17] in Sweden describe a Thermometric Enzyme Linked Immunoabsorbant Assay (TELISA) sensitive to 10^{-13} moles per litre.

Many other sophisticated forms of immunosensor are being constructed commercially and this area is under heavy secrecy and patent cover (e.g. FET and fibre optic devices, ellipsometry, diffraction pattern). However, one of the major, but unfortunately inherent, problems to be solved is the speed of measurement which is often quite slow (e.g. greater than 60 min.). A recent paper from Suzuki's group[26] indicates that the application of an electric pulse of up to 10KHz frequency can achieve 60% antibody-antigen reaction within 5 min. Shorter times are preferable if a truly commercial immuno-biosensor can be realized.

ii. WHOLE CELL BIOSENSORS

Another completely different area of biosensor construction is that using immobilized whole cells or organelles[27,28,29]. Over the last 10 years or so numerous publications have appeared, particularly on bacterial biosensors, mostly from Japanese and American groups. The subject has been recently reviewed[29]. Examples of this type of biosensor that can be used in different commercial areas are given in Table 7, A and B. These sensors are potentiometric or amperometric in design but characteristically have a slow response and often react to a broad spectrum of substrates. The latter property has been taken advantage of in the construction of a microbial biosensor to perform the Ames mutagenesis test in a fraction of the normal assay time[30]. Immobilized Salmonella typhimurium revertants (also Bacillus subtilis Rec) were used in conjunction with an oxygen electrode to measure the mutagenicity and potential carcinogenicity of various chemicals.

TABLE 7(A)

POTENTIAL MICROBIAL BIOSENSORS FOR FERMENTATION AND PROCESS CONTROL

SENSOR FOR:-	MICROBE	RESPONSE TIME (MIN)	TRANSDUCER	STABILITY (DAYS)	RANGE
ASSIMILABLE	BREVIBACTERIUM LACTO-FERMENTUM	1 - 10	O_2	10	LINEAR ABOVE 1MM
ACETATE	TRICHOSPORON BRASSICAE	6 - 10	O_2	21	LINEAR UP TO 22.5 MGL^{-1}
CEPHALOSPORIN	CITROBACTER FREUNDII	10	PH	28	BELOW 50 UG CM^3
ETHANOL	TRICHOSPORON BRASSICAE	10	O_2	21	BELOW 22.5 MGL^{-1}
FORMATE	CLOSTRIDIUM BUTYRICUM	20	FUEL CELL	20	UP TO 1 GL^{-1}
GLUCOSE	PSEUDOMONAS FLUORESCENS	15 - 20	O_2	14	MINIMUM 2MG 1^{-1}

TABLE 7(B)

POTENTIAL MICROBIAL SENSORS FOR ENVIRONMENTAL CONTROL AND GAS MONITORING

SENSOR FOR:-	MICROBE	RESPONSE TIME (MINS)	TRANSDUCER	STABILITY (DAYS)	RANGE
BOD	TRICHOSPORON CUTANEUM	30	O_2	17	1-41 MGL^{-1}
AMMONIA + NO_2	MIXED CULTURE NITRIFYING BACTERIA	4	O_2	10	0.1-42 MGL^{-1}
NITRATE NITRITE HYDROXYLAMINE	AZOBACTER VINELANDII	5 - 10	NH_3	14	10^{-5} TO 8×10^{-4}M
METHANE	METHYLOMONAS FLAGELLATA	1	O_2	20	UP TO 6.6MM
MUTAGEN SCREENING	B. SUBTILIS	90 - 120	O_2	-	1.6 UG CM^3

Whole cell biosensors may come into their own when tailor-made, genetically engineered cells become available to supply a certain enzyme sequence or regenerate complex co-factors, for example. Where these factors are taken into consideration one can see that they may have the edge in cost terms over other forms of biosensor.

7. Practical Biosensors

Practically, we can see roughly four forms of biosensor emerging:

i. Small, hand-held devices. Biochemical matrix can be a membrane, thin probe or dipstick (normally disposable). Electronics - custom designed; readout - digital display to give the answer in seconds. Designed for use by unskilled persons. For example, a device using glucose oxidase in a membrane is being constructed for monitoring diabetes.

ii. The bench-top analyser or multisample autoanalyser. For example, a nylon tube glucose oxidase system used in Technicon auto-analysers[31] is already being used in many laboratories. Many are constructed in-house such as those described by Mascini and Palleschi[32]. They recently made an enzyme nylon coil device for an automated analyser which had a useful life-time of 6 months. One of the few commercially successful bench top analysers uses immobilized enzymes on membranes to measure starch, sugars such as glucose, ethanol, etc. An instrument for mass or automated clinical analyses is also being developed. Here the sample is put on a slide containing entrapped enzymes, buffer and colour reagents. Development of colour in the presence of the appropriate substrate is measured by reflection spectrophotometry. Bench top thermistor devices have already been described.

iii. Flow devices (bioreactors) for 'on line' monitoring of continuous processes, e.g. large volume production in food processing[33], fermentation processes[27] and pollution monitoring[10], and environmental control[28]. An example of this type of biosensor is a portable pesticide monitor, one cubic foot in size and weighing approximately 30 lb. This very neat device employs immobilized cholinesterase on special porous pads through which the suspect water is pumped. If various pesticides, such as organophosphates or carbamates, are present in parts per million, the enzyme is inhibited. The extent of inhibition is measured by injection of a specific substrate and an electrochemical detector. A new pad (they are on a turntable) is then inserted into the sampling flow cell for the next measurement.

Along similar lines an enzyme ChemFET device for detecting nerve gas uses immobilized cholinesterase, but in a dry form.

iv. The continuous in vivo or implanted monitor. This will be a longer term development. Miniaturized implanted devices have been constructed and tested but the lifetime of the sensor is at present too short.

8. Applications and Uses of Biosensors

Biosensors will be widely used in clinical analysis, health care, veterinary and agricultural applications, industrial processing and monitoring, environmental and pollution control. They will have the attraction of being low cost, small, sensitive, selective, rapid and easy to use.

i. CLINICAL CHEMISTRY, MEDICINE AND HEALTHCARE

Benchtop biosensors of the electrochemical variety are of course in current use in clinical biochemistry laboratories for measuring glucose, lactic acid etc. An example of a commercially available instrument is the Yellow Springs Instrument (YSI) Company's glucose analyser which uses a glucose oxidase membrane in conjunction with a modified oxygen electrode. This was originally designed for carbohydrate analysis in the food industry but rapidly gained acceptance in clinical chemistry laboratories. It is expected that other companies will shortly be entering this market with a variety of bench top electrochemical type biosensors.

Another area of clinical medicine and healthcare where commercial biosensors will make an impact is in "out of hours" monitoring. Another example where a hand-held, user friendly type of monitoring is required is in the measurement of blood glucose by diabetics. An insulin dependent diabetic's blood glucose has to be read two to three times a day and it is vital to the health of the individual that such monitoring is accurately performed. Such hand-held devices are already being developed by several companies including Cambridge Life Sciences.

Some examples of biosensor application areas are given below:

Area of Medicine	Characteristics Required
Casualty	Small, single analytes
Bedside special units	
(Intensive care patients)	Benchtop, multiple, invasive
Home care	Hand-held, simplistic
Outpatients	" " "
Drug centres	Small, multiple
Health centres	Benchtop, multiple

Such testing will improve the efficiency of patient care, replacing the often slow and labour intensive present tests. It will bring clinical medicine closer to the bedside, facilitating rapid, clinical decision making. A wide variety of analytes need to be monitored in these situations such as antigens, antibodies, cholesterol, neurochemicals, etc. etc. - the list is legion. In Table 8 we have summarized some examples.

TABLE 8

EXAMPLES OF POTENTIAL BIOSENSOR USES IN DIAGNOSTIC MEDICINE

A	SINGLE TEST - SMALL	
	GLUCOSE	- DIABETES MONITORING
	CHOLESTEROL	- CARDIOVASCULAR
	SPECIFIC DRUG	- COMPLIANCE, ABUSE
B	MULTI-TEST - BENCHTOP	
	GLUCOSE + SPECIFIC IONS, E.G. POTASSIUM	- HEALTH CARE
	CREATINE + UREA	- RENAL FUNCTION
	PROGESTERONE + OESTRONE SULPHATE	- FERTILITY MONITORING
C	INVASIVE - MINIATURE	
	GLUCOSE	- INSULIN PUMP OPERATING
	SPECIFIC DRUG	- THERAPEUTIC DRUG MONITORING
	GLUCOSE, CREATININE, UREA, IONS	- CRITICAL CARE MONITORING

The advent of cheap, user friendly biosensors will revolutionize the practice of healthcare monitoring enabling more in depth studies to be made on a metabolic basis - surely an improvement on the present largely physical tests, for example as in the case of cancer diagnosis/monitoring.

ii. VETERINARY, AGRICULTURE AND FOOD

Here there are many areas where even conventional analysis systems are not available. The introduction of suitable biosenors would have considerable impact in, for instance, the following areas:

Small and large animal care - fertility and infectious disease monitoring

Dairy Industry - Milk (protein, fats, antibodies, hormones, vitamins)

Fruit and Vegetables - viral and fungal diagnosis

Foodstuffs - contamination and toxins, e.g. salmonella

Drink - Brewing - Wine, spirit and beer - improvements in production and Q.C.

iii. FERMENTATION INDUSTRIES, PHARMACEUTICAL PRODUCTION

In addition to alcohol fermentation there are a considerable and increasing number of products being produced by large-scale pro- and eukaryote cell culture. The monitoring of these delicate and expensive processes is essential to reducing and keeping down the costs of production. Additionally, specific biosensors can be designed to measure the generation of a fermentation product.

The use of biosensors in industrial processes will benefit the manufacturer in many ways:

1. A biosensor may be made compatible with both on-line assay and discrete sampling.

2. It provides the opportunity to give a rapid response and, therefore, improved feedback control.

3. It will not interfere with the process stream.

4. A biosensor has a potential lifetime of days, sometimes weeks, which releases technical staff for other duties.

5. It facilitates rapid sampling and rejection of below standard raw materials on delivery.

6. It will provide low cost monitoring of stored products and raw materials.

7. It will give access to remote environments.

8. Biosensors can be made relatively inexpensive.

iv. ENVIRONMENTAL CONTROL AND POLLUTION MONITORING

Because they can be miniaturized and automated, biosensors have many roles to play in these areas. A successful system for pesticide monitoring has already been described.

An area where whole cell biosensors may come into their own is in water monitoring to combat the increasing number of pollutants getting into the groundwater systems, hence into drinking water. There are so many undesirable materials now appearing in the groundwater that a single analyte measurement will be insufficient - a broad spectrum biosensor is required (e.g. see Table 7 - whole cell biosensors). Such a biosensor for BOD measurement is already on the market.

This area of biosensor development is becoming increasingly of interest to the military; for example, as previously mentioned, one company has produced an enzyme biosensor to measure nerve gas. With recent trends towards developing sophisticated biological weapons this area for biosensor development must receive increasing priority.

9. Conclusions

Research into biosensor design and application is still at an early stage in Japan, Europe and the USA. Despite intensive commercial activity experts agree that serious problems remain to be solved. Miniaturization is obviously a key feature and the factors which will be important to the construction of future biosensors will include:

- unique combinations of miniaturized transducers with specific biological materials
- the use of "arrays" of biosensors
- low cost considerations (e.g. adaptation of semiconductor fabrication techniques)
- disposability of devices
- overall, rapid assays achieved
- the ability to perform simultaneous replicate assays increasing the confidence of the accuracy of the assay
- intimate involvement of real time computation - use of computers to "crunch" the data

The evolution of such devices may provide a bonanza for the researcher but it will be some time before ultimate commercial development. However, we are faced with the exciting prospect that such work on the interphase of biological materials with transducers, particularly those which are silicon or fibre optic based, will increase our knowledge of biological molecules and their interaction with electronics to such an extent that ultimately we can contemplate the construction of molecular switching devices to as "biochips" for the computers of the future[34].

REFERENCES

1. M. Gronow, Trends in Biosciences , 1984, 336.

2. C.F.M. Kingdon, D.J. Anderton & M. Gronow, Industrial Biotechnology , Wales, 1985 (in press).

3. O.R. Zaborsky in "Immobilized Enzymes" (I. Chibita, Ed.), CRC Press, 1973.

4. I. Chibita (ed.) " Immobilized Enzymes Research and Development ", Halstead Press, John Wiley & Sons and Kodansha, Tokyo, 1978.

5. K. Venkatsubramanian (ed.) "Immobilized Microbial Cells", ACS Symposium Series 196, American Chemical Society, 1979.

6. G.A. Rechnitz, Science , 1981, 214 , 287.

7. L.C. Clark Jr. and C.H. Lyons, Ann. N.Y.Acad.Sci. (U.S.), 1962, 102 , 29.

8. J.W. Updike and J.P. Hicks, Nature, 1967, 214 , 986.

9. P.W. Carr and L.D. Bowers, "Immobilized Enzymes in Analytical and Clinical Chemistry", John Wiley, New York, 1980.

10. G.G. Guilbalt, Enzyme & Microbial Technol. , 1980, 2 , 258.

11. S. Suzuki, I. Satoh & I. Karube, Appl. Biochem & Biotechnol. , 1982 7 , 147.

12. P. Vadgama, J.Med.Eng. and Tech. , 1981, 5 , 292.

13. C.R.Lowe, Trends in Biotechnol. , 1984, 2 , 59.

14. W.J. Aston and A.P.F. Turner, Biosensors and Biofuel Cells in Biotechnol. and Genetic Eng. Rev., 1984, 1, 89.

15. A.E.G. Carr, G. Davis, G.D. Francis, A.O. Hill, W.J. Aston, J.I. Higgins, E.V. Plotkin, L.D.L. Scott and A.P.F. Turner, Anal. Chem., 1984, 56, 667.

16. S. Caras and J. Janata, Anal. Chem., 1980, 52, 1935.

17. B. Daniellson, B. Mattiasson and K. Mosbach, Pure and Applied Chemistry, 1979, 51, 1443.

18. M.J. Goldfinch and C.R. Lowe, Anal. Biochem., 1984, 138, 430.

19. Y. Ikariyama, M. Furuki and M. Aizawa, "Proc. of an International Meeting on Chemical Sensors", Kodansha, 1983, 693.

20. S. Mansouri and J.S. Schultz, Biotechnology, 1984, 2, (10), 885.

21. M. Aizawa, A. Morioka and S. Suzuki, Anal. Chim. Acta, 1980, 115, 61.

22. Bo Mattiasson and H. Nilsson, Febs. Lett., 1977, 78, 281.

23. J-L. Boitieux, G. Desmet and D. Thomas, Clin.Chem., 1979, 25, 318.

24. T.W. Freeman and W.R. Seitz, Anal. Chem., 1978, 50, 1242.

25. Y. Ikariyama, S. Suzuki and M. Aizawa, Anal. Chim. Acta, 1984, 156, 245.

26. H. Matsuoka, I. Karube and I. Suzuki, Abstracts of Royal Society 'International Symposium on Electroanalysis in Biomedical, Environmental and Industrial Science', UWIST, Cardiff, 1983, Paper No. 11.

27. I. Karube and S. Suzuki, Annual Reports on Fermentation Processes, Academic Press, 1983, 203.

28. H.Y. Neujahr, Biotechnology and Genetic Engineering Reviews, 1984, 1, 167.

29. C.A. Corcoran and G.A. Rechnitz, Trends in Biotechnology, 1985, 3, 92.

30. I. Karube, T. Nakahara, T. Matsunaga and S. Suzuki, Anal. Chem., 1982, 54, 1725.

31. J. Campbell and W.E. Horby in "Biomedical Applications of Immobilized Enzymes and Proteins" (T.M.S. Chang, ed.), Plenum Press, 1977, 3.

32. M. Mascini and C. Palleschi, Anal. Chim. Acta, 1982, 136, 69.

33. I. Karube and S. Suzuki in "Use of Enzymes in Food Technology" (P. Dupry ed.), Technique et Documentation Laboisier, Paris, 1982, 3.

34. J.V. Brunt, Biotechnology, 1985, 3, 209.

14
Enzyme Engineering

By J. M. Walker

DIVISION OF BIOLOGICAL AND ENVIRONMENTAL SCIENCES, THE HATFIELD POLYTECHNIC, P.O. BOX 109, COLLEGE LANE, HATFIELD, HERTS. AL10 9AB, U.K.

1. INTRODUCTION

Industrially used organic catalysts generally lack specificity for the reaction they are catalysing and often have to be used under extreme conditions of temperature and pressure. In comparison, enzymes show high specificity for the reactions that they catalyse and are able to function under extremely mild conditions and at very low concentrations. It is not surprising therefore that enzymes have found a number of applications as catalysts in industrial processes, particularly in the fine chemical, food and pharmaceutical industries and in clinical analysis. However, of the spectrum of enzymes available in nature, only a surprisingly small fraction have as yet found industrial applications. Only about 300 enzymes are commercially available. Of these, only about a dozen have worldwide sales in excess of $10 millions, and these alone account for more than 90% of the total enzymes market (1,2). These figures should be compared with the fact that over 2,000 enzymes have been described in the scientific literature and many thousand more exist in nature although as yet not fully characterised. The reasons for the industrial under-use of enzymes is fairly clear. Enzyme structures have evolved in response to metabolic demands found in vivo, and are consequently well suited to their *in vivo* role. However, when the same enzymes are considered for use as industrial catalysts, they are invariably exposed to unnatural (non-physiological) environments (*e.g.* the presence of organic solvents, elevated temperatures, pH values outside their normal in vivo value *etc.*) which can denature (unfold) the enzyme with consequent loss of activity. One of the major goals of the enzyme technologist therefore is to enhance enzyme stability so that they may function effectively under non-physiological conditions. This will provide enzymes more suitable (*i.e.* with a longer half-life) for the conditions found in industrial processes and should also improve the yield of active enzyme following immobilisation processes. (See chapter by Trevan for a discussion of enzyme immobilisation.) Both the chemical modification of proteins, and the manipulation of solvent media to enhance enzyme stability have found some success (see chapter by Trevan and references 3-6), but the chemical modification of enzymes is harsh and rather non-specific. The recent introduction of the technique of site-directed mutagenesis has provided a far more subtle approach to modifying enzyme structures. This approach, described in detail below (section 2), allows us to specifically alter a chosen base (or bases) in the gene for a given enzyme. This results in the replacement of a specific amino acid by another of one's own choice in the polypeptide chain of the enzyme. Recombinant DNA technology then allows us to produce large quantities of the purified, re-designed enzyme. (See chapter 1 for a discussion of the general procedures for cloning a gene.) Using site-directed mutagenesis we are therefore in a position to alter, in a predictable manner, specific amino acids in an enzyme structure. In this way it should be possible to engineer enzymes with much improved stability characteristics by making appropriate amino acid replacements. These changes will be made based on our understanding of the factors responsible for enzyme stability and a detailed knowledge of the three-dimensional structure of the enzyme.

While enzyme engineering offers us the potential for increasing enzyme stability, many other exciting potential applications are also apparent. These include:

1. The possibility of enhancing catalytic activity and increasing substrate affinity.
2. The modification of substrate specificity, thus constructing novel enzymes from pre-existing enzymes (i.e. creating enzymes not found in nature to catalyse new reactions).
3. Producing an enzyme with an altered pH-activity profile, thus allowing the enzyme to function at non-physiological pH values.
4. The design of enzymes stable to oxidising agents (e.g. washing powder enzymes that will function in the presence

of bleach).

5. The design of enzymes resistant to proteolytic degradation. (Many reaction processes have proteases present, usually as contaminants, which can degrade the enzyme of interest.)
6. Producing enzymes stable and active in non-aqueous solvents.
7. The elimination of allosteric sites involved in feedback inhibition. (The activities of some enzymes are inhibited in the presence of excess product by the product binding to an allosteric site on the enzyme which causes conformational changes resulting in an inactive enzyme. Such effects are undesirable if one is aiming to produce a high yield of product.)
8. The fusion of enzymes involved in a particular reaction pathway, so that a multienzyme process might be carried out using one protein.

By introducing only some of the above improvements, it should be possible to transform many traditional high pressure and high temperature industrial processes into enzymatic processes that can be operated with a low energy consumption. Mutagenesis to produce enzymes with altered activity is of course not new. In the past the search for more suitable enzymes for industrial use has included extensive searches for improved enzymes by mutation and selection programmes to enhance the properties of the native enzyme. The random mutation of microorganisms to produce strains with enhanced characteristics is a well tried technique, particularly within the pharmaceutical industry. A classic example of this general approach is the bacterial production of penicillin which has been improved about 10,000-fold over the last 40 years. There are many cases where the activities of bacterial enzymes have been directed toward novel substrates in vivo by selecting spontaneous mutants of the enzyme which allows the organism to grow on the new substrate (7-10). Enzymes lacking allosteric inhibition have also been isolated using this approach (11). However, these conventional mutagenesis techniques are generally limited to producing very minor (usually single amino acid) changes in the enzyme structure. Should it prove necessary to change several specific amino acids throughout the protein chain to make a particular improvement, one is unlikely to detect such an event in a mutant population because of the extremely small probability of its occurrence. The introduction of such multiple alterations <u>is</u> possible, however, using site directed mutagenesis. Also screening for random mutations is a highly time-consuming procedure which lacks the subtlety and design aspect provided by site-directed mutagenesis. However, such mutation studies have shown us that enzyme properties can indeed by improved, and it is anticipated that enzyme engineering by site directed mutagenesis can build on this knowledge to provide even greater improvements. The ultimate goal of enzyme engineering must be the design and construction of enzymes to order such that they will catalyse <u>any</u> desired reaction, if necessary even under the most extreme conditions. Although the potential for enzyme engineering is enormous, at present our ability to apply the technique to its full potential is limited by a number of factors. These include the following.

1) A lack of x-ray crystallographic data for many enzymes means that the three dimensional structure of the majority of enzymes is not yet known. Such structural data are essential for each enzyme to be modified, since they provide the structural basis upon which appropriate changes in enzyme structure are decided. For those enzymes where the three dimensional structures are presently available, the application of computer graphics, which depict the translation and rotation of an enzyme and its substrate, in real-time, is providing a powerful tool in helping to design appropriate changes to enzyme structure (12-15, 37). Some of the problems involved in determining the three-dimensional structures of proteins are described in reference 2. This lack of detail on the three-dimensional structures of enzymes is likely to prove the major rate-limiting step in the progress of enzyme engineering over the next few years.
2) We lack the knowledge of the exact interactions (hydrogen bonding, electrostatic interactions and hydrophobic interactions) made between an enzyme and its substrate at the active site. Without this information it is not possible to predict the effect of altering specific amino acids at the active site. Many of the initial experiments in enzyme engineering have therefore involved investigating the effect of removing residues suspected of being involved with

substrate binding and observing the consequent effect on substrate binding (see section 3).

3) We lack knowledge of the precise factors involved in conferring stability to proteins. However, it is clear that salt bridges and other electrostatic interactions confer thermostability, as do amino acid changes that stabilise secondary structures and interactions between secondary structures. Again, initial experiments in enzyme engineering have involved the investigation of the effect on stability of substituting residues thought to be involved in enzyme stability (see section 3).

Already a number of exciting examples of enzyme engineering have appeared in the literature. Given the above limitations, it is not surprising that initial studies have been made on enzymes with a known or suspected relationship between structure and function. Some examples of the recent experiments in enzyme engineering are described in section 3 and provide a good indication of the potential of this technique. Although this article is restricted to a description of enzyme engineering, it should be stressed that the basic approach is equally applicable to the improvement of the stability or function of any protein, e.g. clinically useful compounds such as interferon or growth hormone, veterinary and agricultural products, etc. References to proteins that have been modified by site directed mutagenesis are given at the end of section 3.

2. SITE-DIRECTED MUTAGENESIS

The term site-directed mutagenesis refers to any technique that allows one to specifically (site-directed) change (mutate) a base in a length of DNA. There are in fact a number of different technical approaches to achieving this end, and these have been well reviewed in references 16-18. However, one technique in particular, involving the use of synthetic oligonucleotides, is finding particular use in enzyme engineering and is described here. It should be stressed that all the site-directed mutagenesis techniques require the cloning of the gene for the enzyme under study and its incorporation into a suitable carrier such as a plasmid or bacteriophage vector. This, however, should not be a problem since genetic engineering technology has now reached the point where we can now clone the gene for essentially any protein found in nature. (See chapter 1 for a description of the general procedures involved in cloning a gene.)

The technique of oligonucleotide-directed mutagenesis requires the synthesis of a short (15-20) oligonucleotide that is complementary (i.e. base pairs) to the gene around the site to be mutated, but which contains a mismatch at the base that we wish to mutate. This of course requires that the base sequence of the enzyme gene has been determined, but this is not a difficult task nowadays. The methodology for the synthesis of moderately long oligonucleotides has rapidly developed in recent years and nowadays such molecules can be synthesised chemically in a matter of a day or so. The basic principle is very simple. For two nucleotides to be joined the 5' end of one and the 3' end of the other are blocked. Two different types of blocking groups are used; one that can be removed by acid and one with base. One can then chemically condense a 5' blocked mononucleotide with a 3' blocked molecule to give a dinucleotide blocked at both ends. Selective removal of either the 5' or 3' blocking group (using either acid or base) allows the dinucleotide to react with an appropriately unblocked mono- or dinucleotide. This cycle can then be repeated many times to provide the oligonucleotide of the desired length. A further development has been the use of solid supports when the first nucleotide is attached to the support and then further nucleotides added in stepwise fashion, washing the support between each step. This approach has been automated, resulting in the production of 'gene machines' for the synthesis of oligonucleotides. A detailed description of the chemistry of oligonucleotide synthesis is given in an excellent review by Itakura (19).

The basic procedure for oligonucleotide-directed mutagenesis is shown in Figure 1. We start with a single-stranded clone of the complementary strand of the enzyme molecule, carried in an M13 phage vector (see chapter 1). This is then mixed with the synthetic oligonucleotide. Although there is a mismatch, as long as this mismatch is near the centre of the oligonucleotide, and as long as the mixing is done at low temperature in the presence of high salt concentration, the oligonucleotide will hybridise (bind) to the appropriate position on the enzyme gene. DNA polymerase is then introduced and uses the oligonucleotide

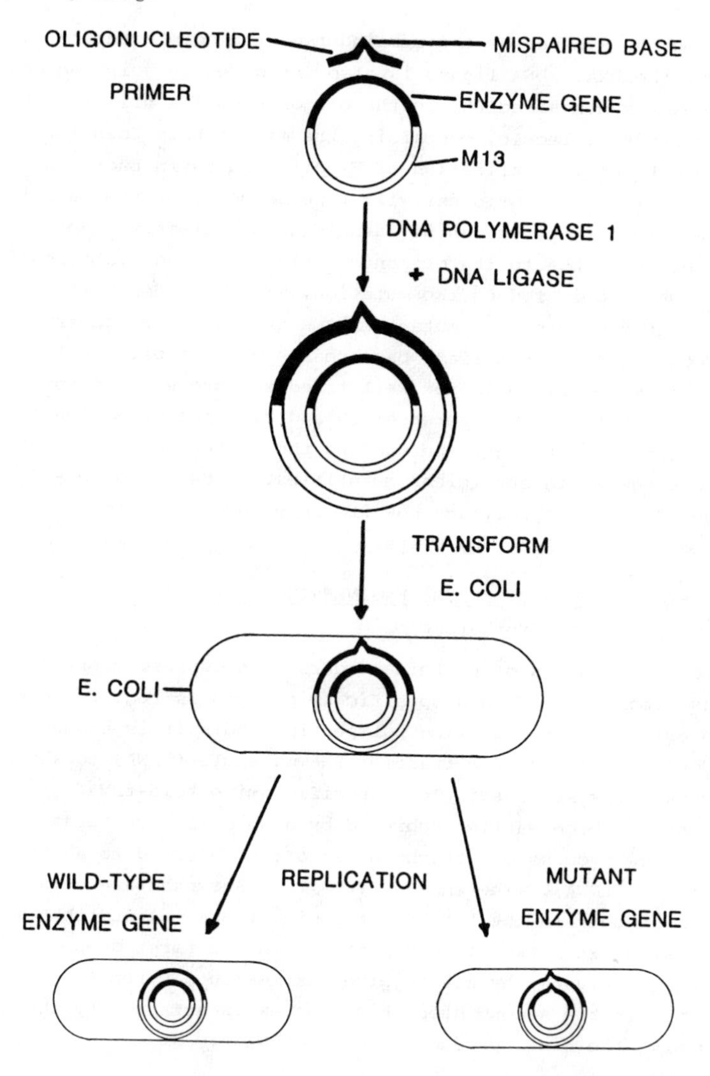

FIGURE 1: Site-directed mutagenesis using a synthetic oligonucleotide (see text for details)

as a primer to synthesise the remainder of the complementary strand of the DNA. DNA ligase is also intrduced to join the ends of the newly synthesised DNA to the oligonucleotide primer. This double stranded molecule, containing the mismatch, is then introduced into E.coli. Replication of E.coli results in bacteria containing either the original wild type sequence or the mutant sequence. Mutant clones can be identified by selective hybridisation of their DNA to the oligonucleotide (now radiolabelled) that was used to introduce the mutation. These clones will therefore be producing the mutant enzyme that we have constructed by introducing the appropriate base change *via* the oligonucleotide primer. This technique is now well tried and proven, and further details can be found in references 20 and 21. The only significant problem encountered is persuading the bacterium to produce the mutant enzyme in sufficient quantities. A detailed description of this problem, and how it can be overcome, is described in the chapter by Slater.

3. SPECIFIC EXAMPLES OF ENZYME ENGINEERING

(a) Tyrosyl tRNA Synthetase

In the cell, synthesis of proteins occurs on cytoplasmic structures known as ribosomes. When a specific amino acid is required for incorporation into the growing polypeptide chain it is brought to the ribosome by a specific transfer RNA molecule (tRNA) to which it is covalently attached. This specific amino acid-tRNA attachment has been earlier achieved by a specific synthetase enzyme. (The process of attachment is often referred to as the 'charging' of tRNA). The enzyme tyrosyl tRNA synthetase catalyses the reaction between the tRNA for tyrosine ($tRNA^{tyr}$) and the amino acid tyrosine, in a two-stage reaction. In the first step tyrosine is activated by ATP to give enzyme-bound tyrosyl adenylate. In the second step this complex is attached by the tRNA to give the final product tyrosine-tRNA.

(1) Tyrosine + ATP $\rightleftharpoons$ tyrosyl adenylate + pyrophosphate
(2) Tyrosyl adenylate + $tRNA^{tyr}$ $\rightleftharpoons$ tyrosine-tRNA + AMP

The enzyme tyrosyl tRNA synthetase from B.stearothermophilus has been purified, crystallised, its three-dimensional structure determined by x-ray crystallography and the active site identified (22,23). With the aid of computer graphics, the possible effects of changing amino acid side-chain contacts from the enzyme to the substrate, or of distorting the polypeptide chain at the active site have been predicted, and then these theoretical predictions have been tested in practice using site directed mutagenesis. This enzyme probably represents the most detailed study to date of enzyme engineering of a given enzyme (24-27, 37). The following summarises the information that has been obtained by site-directed mutagenesis of this enzyme.

(1) X-ray studies showed that the enzyme residues cysteine-35, histidine-48 and threonine-51 all appear to form hydrogen bonds at the active site with the ribose moiety of tyrosine adenylate. To confirm the existence of these contacts, and to determine their relative importance, a series of mutants were constructed in which histidine-48 was altered to glycine, threonine-51 to alanine and cysteine-35 to glycine respectively. Each of these changes eliminates the particular hydrogen bond involved. From kinetic data on each of these mutants it was deduced that the imidazole side chain of histidine-48, and the sulphydryl side chain of cystine-35 were indeed involved in binding to the substrate, each contributing about 1 kcal/mole to the stability of the transition state in tyrosine activation. (For example, the replacement of cysteine-35 with glycine reduced the catalytic activity of the enzyme by about 70%, mainly by lowering the strength of tyrosyladenylate binding.) Surprisingly, removal of the hydrogen bonding by threonine-51 actually led to an <u>increase</u> in the binding of substrate by about a factor of two (26). The exact reason for this increased binding has yet to be ascertained, but a possible explanation has been postulated (see reference 27). The above observations do however confirm the feasibility of using site directed mutagenesis to examine predictions made from structural studies on the active site of an enzyme.

(2) Comparison of the structure of tyrosyl-tRNA synthetase from <u>B.stearothermophilus</u> and <u>E.coli</u> showed that threonine-51 was replaced by proline in the <u>E.coli</u> enzyme. This was surprising, since the presence of proline should disrupt the α-helical structure of the polypeptide backbone found in this region. A mutant of the <u>B.stearothermophilus</u> enzyme

was therefore constructed with proline at residue 51 (26). Although this would indeed have disrupted the backbone of the enzyme, the binding of the enzyme to ATP was surprisingly found to have increased by a factor of 100. The construction of a double mutant (27) was used to show that the presence of proline-51 caused its effect by strengthening the contact between histidine-48 and the substrate, presumably by inducing a conformational change in the region of histidine-48 resulting in the imidazole side-chain being better placed to form a stronger hydrogen bond with the substrate. This double mutant (glycine-48, proline-51) had low affinity for the substrate, indicating that both the imidazole side-chain of histidine-48 and the proline-51 mutation are necessary for the increased affinity for ATP. Further, it was shown that histidine-48 contributed 3 kcal/mol to transition-state binding in the proline-51 enzyme, but only 1.2 kcal/mol in the wild-type of enzyme. The importance of these observations is that they show that enzyme affinities <u>can</u> be improved by <u>in vitro</u> manipulation, an observation which has important consequences for the commercial applications of enzyme engineering.

(3) In the tyrosine binding-site of the enzyme, the phenolic hydroxyl of the tyrosine substrate appears to hydrogen bond to the side-chains of tyrosine-34 and aspartic acid-176 in the active site 'pocket'. The enzyme shows high specificity for the substrate tyrosine. The enzyme discriminates phenylalanine, which also has an aromatic ring side-chain, from tyrosine by a factor of 10,000 in the charging of $tRNA^{tyr}$. The phenolic hydroxyl of tyrosine-34 seems to be a major determinant in discriminating against the binding of phenylalanine at the active site. However, when tyrosine-34 was replaced by phenylalanine, relatively little effect was observed on the binding of tyrosine but a fifteen-fold decrease in the discrimination against phenylalanine was observed (27). Although this is only a relatively small change, it does show that site-directed mutagenesis offers the potential to alter the <u>specificity</u> of an enzyme.

(b) β-lactamase

β-lactamase is an enzyme produced by certain antibiotic resistant bacteria. The enzyme catalyses the hydrolysis of the amide bond of the lactam ring of penicillins and related antibiotics (e.g. cephalosporins) and thus confers resistance to these antibiotics in the bacterium. The catalytic pathway includes an acyl intermediate and residues serine-70 and threonine-71 have been implicated as being important residues at the active site of the enzyme. It has been suggested that the hydroxyl group of serine-70 adds nucleophilically to the carbonyl group of the β-lactam ring, in a mechanism somewhat analogous to that of serine proteases. The role of threonine-71 is less clear, but it seems essential for catalytic activity because a mutant (not genetically engineered), probably with isoleucine at this position, shows no catalytic activity. The role of this region of the molecule in the catalytic activity of the enzyme has been investigated by site directed mutagenesis (21,28,29).

(1) The conversion of serine-70 to threonine gave a product with no β-lactamase activity. This was not surprising since the conversion has the effect of adding a methyl group to the serine side-chain, thereby hindering access to the hydroxyl group. However, the role of serine-70 in the catalytic activity of the enzyme has been confirmed by this experiment.

(2) The serine-70 residue in the wild-type enzyme has been replaced by a cysteine residue (28,29). This effectively replaces one nucleophilic group (-OH) by another somewhat bulkier one (-SH) producing a thiol β-lactamase. This thiol enzyme has a binding constant for penicillin about the same as that for the wild-type enzyme, but the rate of hydrolysis is only about 1-2% of that of the wild-type enzyme. However, against certain cephalosporin antibiotics (which also contain β-lactam rings) the binding constant is more than ten-fold greater than that of the wild-type enzyme and the rate of hydrolysis at least as great as that of the wild-type enzyme. The mutation therefore has essentially changed the specificity of the enzyme. Additionally, the mutant enzyme is three times more resistant to trypsin digestion than the wild-type enzyme. This difference has been related to increased thermal

stability of the mutant enzyme, one of the major goals of enzyme engineering.

(c) Dihydrofolate Reductase

Dihydrofolate reductase (DHFR) catalyses the reduction of 7,8-dihydrofolate to 5,6,7,8-tetrahydrofolate, which in turn has a major metabolic role as a carrier of one-carbon units in the biosynthesis of purines, thymidylate and some amino acids. The inhibition of DHFR by synthetic folate analogues (antifolates) results in the depletion of the cellular tetrahydrofolate pool, with consequent cessation of DNA synthesis leading to stasis and cell death. Such an approach is especially useful in destroying dividing cells (such as tumour cells) which are active in DNA synthesis. Antifolates, especially methotrexate, have therefore found particular use in the chemotherapy of cancer. It is not surprising therefore that DHFR is an enzyme of considerable interest to pharmaceutical chemists and drug designers. The three dimensional structure of DHFR has been determined (30) and site-directed mutagenesis has recently been used to answer a number of questions concerning the structure and function of DHFR (31,37).

(1) The crystal structure suggests that aspartic acid-27, buried below the enzyme surface, may form a hydrogen-bonded salt linkage with the pteridine ring of the substrate. It would appear to ultimately act as a proton donor and to stabilise the transition state by providing a negatively-charged carboxylate counter ion to hydrogen bond with the resulting positively-charged pteridine ring. To test this, aspartic acid-27 was replaced by asparagine, which leaves the geometry of the site unchanged, but eliminates hydrogen bonding and formation of a cation. The mutant enzyme had only 0.1% of the activity of the wild-type enzyme, supporting the postulated role of the aspartic acid-27 side-chain in catalysis.

(2) The crystal structure suggests that glutamic acid-139 may form a salt bridge with histidine-141, which may stabilise the β-sheet structure in that region. Glutamic acid-139 was therefore replaced with lysine which would not form a salt bridge. This alteration did not affect the catalytic

activity of the enzyme, but introduced a significant decrease in stability of the enzyme. This is an important observation since it shows that structural stability can be uncoupled from enzyme activity.

(3) DHFR contains two cysteine residues but these are not involved in the formation of a disulphide bridge. The crystal structure suggested that if proline-39 was replaced by a cysteine residue it could form a disulphide bridge with cysteine-85 which might improve the stability of the enzyme. When this change was made, the reduced form of the enzyme (no disulphide bridges) showed normal activity. However, when oxidised to form a disulphide, enzymic activity was significantly reduced, presumably due to a loss of dynamic flexibility in the molecule. Obviously in this particular experiment the attempt to improve the stability of the enzyme was detrimental to enzyme activity.

(4) All the DHFR enzymes studied to date (from a variety of sources) contain a glycine-glycine dyad at positions 95 and 96, the glycine residues being linked by an unusual *cis* peptide bond. It is thought that this particular topography may play a role in the working of the molecule, possibly as a conformational switch of some kind. A minimal change which should nevertheless alter the geometry in this part of the molecule would be the replacement of glycine-95 with alanine. When this mutation was made the enzyme was found to be completely inactive, and the mutant enzyme also had a lower mobility on non-denaturing gels, suggesting that a change in the conformation of the enzyme had occurred. The essential nature of the glycine-glycine dyad for the functioning of DHFR has therefore been confirmed by this experiment.

The above examples have been presented to indicate the scope offered by site-directed mutagenesis both to examine the relationship between enzyme structure and function, and to improve this relationship as well as improving enzyme stability. The technique of enzyme engineering is obviously still in its infancy, but one can look forward to an increasing number of publications in this area over the next few years. Indeed, the above examples only represent a part of the papers on enzyme and protein engineering published at the time of writing. Lack of space precludes the description of further examples. However, the following list is a collection of some further enzymes and proteins that have been investigated by site-directed mutagenesis and directs the interested reader to appropriate references.

$\underline{\alpha_1}$-antitrypsin: Mutants with therapeutic potential have been constructed as well as mutants with specificity towards different proteolytic enzymes (32,38).

Insulin: Mutants of the interstitial C-peptide have been constructed to investigate the roles of these regions (33).

Interleukin-2: The role of the cysteine residues in the functioning of the protein have been investigated by site-directed mutagenesis (34).

Carboxypeptidase A and Trypsin: The role of active-site residues have been investigated (35).

Lysozyme: A disulphide bridge has been introduced into the enzyme producing a thermally more stable enzyme with unaltered catalytic activity (36).

Triosephosphate isomerase: Changes in free-energy profiles have been determined following changes to residues involved in the catalytic activity of the enzyme (37).

Dihydrolipoamide acetyl transferase (part of pyruvate dehydrogenase multienzyme complex): The effect of deletions and mutations on the assembly, catalytic activity and active site coupling in the complex have been investigated (37).

Subtilisin: The effects of mutations on the specificity and susceptibility to oxidation of the enzyme have been investigated (37).

Monoclonal antibodies: Site-directed mutagenesis has been used to analyse the structural basis of the interactions of antibodies with their antigens or their effector proteins (37).

References

1. Katchalski-Katzir, E. and Freeman, A. (1982) Trends. Biochem.Sci., 7, 427.

2. Ulmer, K.M. (1983) Science, 219, 666.

3. Noet, K.E., Nanci, A. and Koshland, D.E.Jr. (1968) J.Biol.Chem., 243, 6392.

4. Kowalski, D. and Laskowski, M. (1976) Biochemistry, 15, 1300 and 1309.

5. Freeman, A. (1984) Trends in Biotechnology, Vol.2, No. 6., 147.

6. Means, G.E. and Feeney, R.E. (1971) Chemical Modification of Proteins. Pub. Holden-Day.

7. Hall, B.G. (1981) Biochemistry, 20, 4042.

8. Turberville, C. and Clarke, P.H. (1981) FEMS Microbiol. Lett., 10, 87.

9. Hall, A. and Knowles, J.R. (1976) Nature, 264, 803.

10. Scazzocchio, C. and Sealy-Lewis, H.M. (1978) Eur.J.Biochem., 91, 99.

11. Pabst, M., Kuhn, J. and Sommerville, R. (1973) J.Biol.Chem., 248, 901.

12. Langridge, R., Ferrin, T., Kuntz, I. and Connolly, M. (1981) Science, 211, 661.

13. Richards, G. and Sackwild, V. (1982) Chem.Br., 18, 635.

14. Pensack, D.A. (1983) Industrial Research and Development. Jan. p.74.

15. Meltzer, R.S. and Freeman, J.T. (1983) Computer Graphics News, Jan./Feb. p.3.

16. Smith, M. (1982) Trends in Biochemical Sciences, Dec. p.440.

17. Shortle, D., Di Maio, D. and Nathans, D. (1981) Ann.Rev. Genetics, 15, 265.

18. Timmis, K.N. (1981) in Society for General Microbiology Symposium (ed. Glover, S.W. and Hopwood, D.A.) vol.31, p.49.

19. Itakura, K. (1982) Trends in Biochemical Sciences, Dec. p.442.

20. Zoller, M.J. and Smith, M. (1982) Nucleic Acids Research, 10, No. 20, 6487.

21. Dalbadie-McFarland, G. et al. (1982) Proc.Natl.Acad.Sci., 79, 6409.

22. Bhat, T., Blow, D., Brick, P. and Nyborg, J. (1982) J. Molec.Biol., 158, 699.

23. Rubin, J. and Blow, D. (1981) J.Molec.Biol., 145, 489.

24. Wilkinson, A.J., Fersht, A.R., Blow, D.M. and Winter, G. (1983) Biochemistry, 22, 3581.

25. Winter, G., Fersht, A.R., Wilkinson, A., Zoller, M. and Smith, M. (1982) Nature, 299, 756.

26. Wilkinson, A.J., Fersht, A.R., Blow, D.M., Carter, P. and Winter, G. (1984) Nature, 307, 187.

27. Winter, G. and Fersht, A.R. (1984) Trends in Biotechnology, Vol.2, No. 5, p.115.

28. Sigal, I.S. et al. (1984) J.Biol.Chem., 259, 5327.

29. Sigal, I.S., Harwood, B. and Arentzen, R. (1982) Proc. Natl.Acad.Sci., 79, 7157.

30. Filman, D., Bolin, J., Matthews, D. and Kraut, J. (1982) J.Biol.Chem., 257, 13663.

31. Villafranca, J. et al. (1983) Science, 222, 782.

32. Courtney, M. et al. (1985) Nature, 313, 149.

33. Wetzel, R. et al. (1981) Gene, 16, 63.

34. Wang A. et al. (1984) Science, 29 June, 1431.

35. Kaiser, E.T. (1985) Nature, 313, 630.

36. Perry, L.J. and Wetzel, R. (1984) Science, 226, 555.

37. In: Design, Construction and Properties of Novel Protein Molecules. Proceedings of a Royal Society Discussion Meeting held on 5-6 June 1985, organised by Professor D.M. Blow, Professor A.R. Fersht and Dr. G. Winter. Pub: Royal Society, London (in press).